Native Florida Plants for Drought- and Salt-Tolerant Landscaping

Richard Wunderlin and George R. Kish

Sparkleberry (see pages 68–69)

Pineapple Press, Inc.
Sarasota, Florida

To the memory of Betty Wargo (1932–2013), longtime member of the Suncoast Native Plant Society, who generously contributed her time and numerous plant photographs for use by the society and for the USF Atlas of Florida Vascular Plants.

Inquiries should be addressed to:

Pineapple Press, Inc.
P.O. Box 3889
Sarasota, Florida 34230

www.pineapplepress.com

Library of Congress Cataloging-in-Publication Data

Wunderlin, Richard P., 1939–
Native Florida plants for drought- and salt-tolerant landscaping / Richard Wunderlin and George R Kish.
pages cm
Summary: "Descriptions and photographs for 70 native plants that will thrive with little care in the yards of most Florida homeowners. Covers the peninsula south of Marion, Levy, and Volusia counties through the Keys" — Provided by publisher.
Summary: "Florida is known throughout the world for its beautiful beaches and coastal areas. Not surprisingly, the majority of Florida's population lives near the coast. Gardening in coastal and dry inland areas can be quite challenging. This book will introduce you to native plants that have adapted over time to nature's harshest conditions: heat, humidity, insects, storms, floods, freezes, frost, salt spray, and drought" — Provided by publisher.
Includes bibliographical references and index.
ISBN 978-1-56164-560-2 (pbk.)
1. Endemic plants—Florida. 2. Drought-tolerant plants—Florida. 3. Landscape gardening—Florida. I. Kish, George R. II. Title.
QK154.W866 2013
712.09759—dc23

2012035446

First Edition
10 9 8 7 6 5 4 3 2 1

Design by Shé Hicks
Printed in the United States

Map of Peninsular Florida Counties

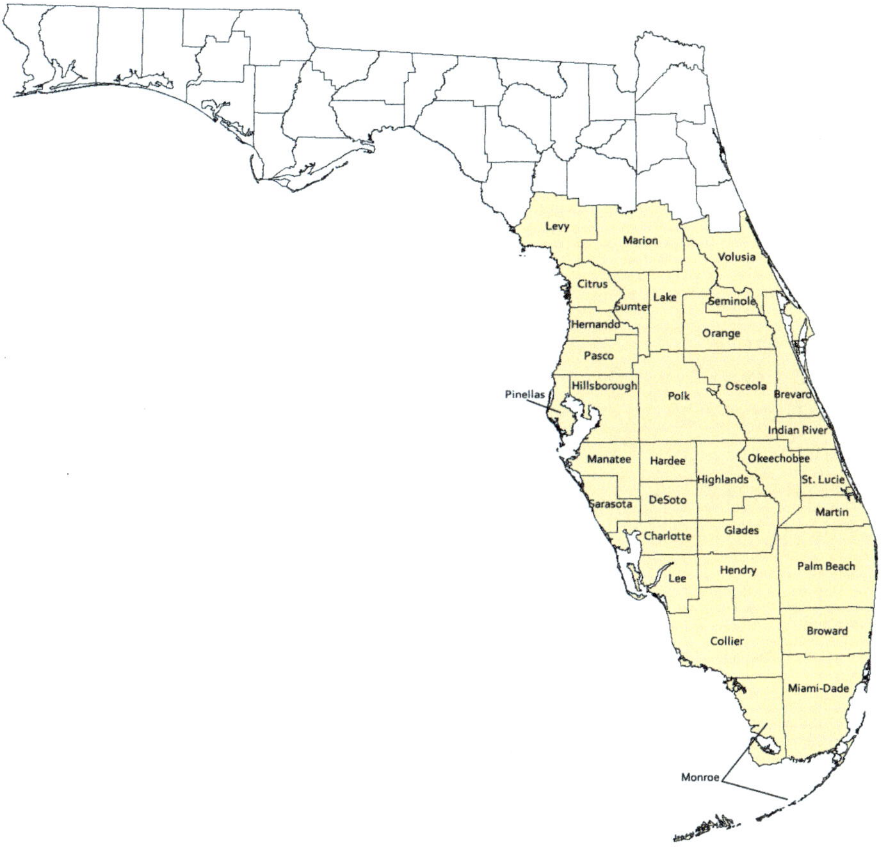

CONTENTS

ABOUT THE FLORIDA NATIVE PLANT SOCIETY

The Florida Native Plant Society is a not-for-profit corporation whose purpose is to promote the preservation, conservation, and restoration of the native plants and native plant communities of Florida. FNPS members throughout Florida are organized into 39 area chapters, and the Suncoast Native Plant Society is one of those chapters. Membership meetings, field trips, plant sales, and other activities are scheduled by individual chapters and usually are open to the public. For the name of the FNPS chapter nearest you, write the Florida Native Plant Society, P.O. Box 278, Melbourne, FL 32969-0278 or visit **www.fnps.org**. For information about the Suncoast Native Plant Society, write the society at P.O. Box 82893, Tampa, FL 33682-2893 or visit **www.suncoast.org**.

INTRODUCTION

Florida is known throughout the world for its beautiful beaches and coastal areas. Not surprisingly, the majority of Florida's population lives near the coast. Gardening in coastal and dry inland areas can be quite challenging. This book will introduce you to native plants and landscaping for these areas in peninsular Florida: Marion, Levy, and Volusia Counties south through the Keys.

Why Native Plants Are Right for You

If you plant a variety of natives, your landscape will come alive with the lovely sights, floral fragrances, melodious birds and insects, and varied wildlife activities of our unique Florida heritage.

Used in suitable locations, native plants are easy and fun to grow and require exceptionally little maintenance. They have adapted over time to nature's harshest conditions—heat, humidity, insects, storms, floods, freeze, frost, salt spray, and drought. These plants are tolerant of peninsular Florida's dry seasons and hot, wet summers. Furthermore, native wildflowers commonly reseed, coming back year after year on their own. Once your landscape is established, you can relax and enjoy it with only minor maintenance.

You help preserve the environment by using native plants. You conserve natural resources, such as water and fossil fuels, and also provide food and shelter for native wildlife.

What Is a Florida Native Plant?

For most purposes, "Florida Native Plant" refers to those species that occurred in Florida prior to the arrival of the Europeans. However, a precise definition is a subject of debate among botany experts. For a discussion of the challenges with determining a technical definition of "native," visit the Florida Native Plant Society website (www.fnps.org).

What Is Native Plant Landscaping?

Native plant landscaping is based on the principle of putting the right native plant in the right place. Rather than first choosing your plants and then modifying the site to suit them (a process that often requires much time, effort, topsoil, fertilizer, pesticide, water, and mulch), you first determine the nature of the site and then select native plants that will thrive in that location. An easy way to make your selection is to choose native plants that grow in natural habitats similar to your site. Once established in the right place, your plants will require very little maintenance. Even though there are some differences between soils in the wild and urban soils that have been disturbed, many native plants, including the ones in this book, do extremely well on urban plots.

Who Needs Salt-Tolerant Plants for Dry Areas?

Anyone living on the east or west coast of peninsular Florida may benefit from using salt-tolerant plantings. Coastal plantings must be tolerant of the alkaline, shelly, salty, xeric (droughty) sands typical of coastal areas. These areas can also receive aerial salt spray, occasional storm-induced saltwater inundation, and even regular tidal inundation. It is important to know which plants can tolerate these harsh conditions.

Inland coastal areas may have brackish irrigation water sources due to saltwater intrusion. Reclaimed water, an important source of irrigation water, provides an economical alternative to potable water in many areas. However, it may have elevated chloride levels.

As you browse through this book, you will note that most of the described plants tolerate full sun and dry soils (once established). So whether you live on the coast, near it, or even farther inland, this book may help you choose appropriate plants for your peninsular Florida landscape and increase the plant diversity and habitat value.

Cold-Sensitive Plants

All of the plants listed and described in this book are native to peninsular Florida. Some are native to the warmer coastal areas and are not hardy in the colder inland areas. However, they may thrive with special protection and care during cold events. Some of these plants will freeze to the ground but may emerge from the roots the following spring depending on the severity of the freeze, establishment of the plant, and special protection. Some tender plants may also reseed in the spring. Please observe the range requirements noted in this book or implement frost protection measures for cold-sensitive plants.

How to Use This Book

Start with Basics of Native Plant Landscaping for Dry and Coastal Sites on pages 5–9. This section includes much information that is specific to native plant landscaping, so please read it even if you are experienced in other, more traditional landscaping methods. Then use the plant descriptions section on pages 11–127 to select your plants. Note that this section is divided into three parts: "Trees," "Shrubs," and "Groundcovers, Vines, and Wildflowers." The Quick Reference Guide on pages 128–135 can help you do this more quickly. If you are interested in a certain characteristic, such as a specific height at maturity or the ability to tolerate shade, you can scan the Quick Reference Guide, find plants with the desired characteristic, then refer back to the plant descriptions for more detail.

We selected plants in this book for ease of growing and availability. Unfortunately, we had to omit many excellent plants for lack of space. A few additional species that may be of interest to you are listed on page 136. Your local chapter of the Florida Native Plant Society can recommend other desirable plants. If you cannot find a desired plant for sale, ask for it, as many nurseries will order plants on request. See Native Plant Sources under Resources on page 138 for more information on this topic.

For descriptions of the three major dry coastal plant communities (beach dune, coastal strand, maritime forest), read Dry Peninsular

Florida Plant Communities on pages139–140. If you live in a coastal area, your homesite is most likely situated in one of these plant communities. Each plant listed in the plant descriptions naturally inhabits a number of different communities. Most plants listed occur in at least one of the three major dry coastal habitats.

If you are tempted to gather from the wild rather than to purchase container-grown plants, please read Florida Native Plant Society Policy on Transplanting Native Plants from the Wild on pages 141–142. You should also be aware that collecting plant material such as seeds, cuttings, leaves, or whole plants from property that you do not own may violate Florida Statute 581.185, "Preservation of Native Flora of Florida."

Resources on pages 137–138 lists plant sources, helpful organizations, and places to view native plant landscapes and natural habitats.

The Bibliography on pages 143–146 offers a selection of helpful publications.

BASICS OF NATIVE PLANT LANDSCAPING FOR DRY AND COASTAL SITES

Here are the basic principles and methods for planning, planting, establishment, and maintenance of native plant landscaping.

PLANNING

Native plant landscaping begins with a decision to preserve natives that are already present. The next step is to understand the unusual conditions of dry sites. Your plan may be as ambitious as an all-native landscape for the entire site or as simple as a small plot of wildflowers.

Preservation of natives. If you are not sure what native plants you already have, your local Native Plant Society chapter can help with identification. If you are involved in construction of a building, protect the native vegetation, including root zones, from fill, digging, scraping, and damage from machinery. Ask the Cooperative Extension Service of the county where you plan to build for information on plant protection during construction.

Soils. Coastal soils usually are highly alkaline and contain numerous shells. These conditions limit the choices of plants to those that can tolerate the high alkalinity. Beach strand and dune soils generally lack organic matter while maritime forest soils may have an organic layer. Your county Cooperative Extension Service will be able to help you determine the types of soils in the various parts of your site.

Light. Most dry habitat plants bloom more profusely in full sun, but some also do well in partial shade as noted in the individual plant descriptions.

Spacing. Because dry habitat plants need large root systems to use what water is available, it is natural for these plants to be spaced farther apart than wet habitat plants. Planting too closely may result in a crowded, weedy-looking landscape that requires thinning and pruning. The plant descriptions tell expected sizes for mature plants. Space far enough apart to allow for their full growth.

Wildlife. For wildlife cover, you need a variety of plants of different heights: groundcover, vines, shrubs, and trees. Include some evergreen plants for winter shelter and birdhouses or nest boxes for birds.

A varied selection of native plants provides seeds, berries, nectar, and leaves for wildlife food. In addition to cover and food, be sure to provide a water supply, such as a birdbath.

The larger the area containing desirable plants, the more likely it is to attract wildlife. For this reason, neighbors sometimes plan wildlife gardens together. Such planning is an exciting feature of the Florida Yards and Neighborhoods Program, offered by your Cooperative Extension Service.

For more information about attracting wildlife, see the Bibliography.

PLANTING

Now that you have selected appropriate plants and decided where to put them, it is time to plant. Dig your hole just deep enough that the flat top of the root ball is level with the ground outside the hole. It is important not to plant too deeply or too shallowly. If the plant is pot bound, carefully loosen the roots and direct them into the soil so they can establish a good root system. After putting a plant into the hole, fill the hole with water and put the soil back around the root ball, forming it with your hands. If the water drains so fast that you cannot fill the hole, keep adding water as you add soil. Starting with water in the hole eliminates air pockets around the roots and gives the plant a thorough initial watering.

ESTABLISHMENT

A plant is considered established when it is able to get along without supplemental watering except in times of serious drought. Establishment of plants in a dry landscape requires frequent watering at first, followed by a tapering-off period. Depending on the plant, its size, and the season, establishment can take considerable time—for some plants, more than a year. Growth will be slow until roots are established. In general, large plants take longer to establish than small ones, and all plants take longer during hot, dry seasons.

When you buy a plant, ask how long establishment takes, how frequently to water while the plant is still in the pot, and what watering schedule to use after the plant is in the ground. Check new plants frequently and water them if they wilt, but outside of that, do not give water in excess of the recommended schedule. Dry habitat plants, even in containers, are quite vulnerable to damage by overwatering.

MAINTENANCE

Once a native plant is established in the right place, it requires very little maintenance. Basic maintenance includes water, fertilizer, mulch, and managing pests.

Water. Native plants established in the proper habitat generally do not require supplemental watering. However, in times of drought, watch your plants carefully for signs of wilting, yellowing, or lack of vigor, and water any struggling plants.

It is best to check for wilting in the early morning, since a plant that does not need water may still wilt in the heat of the day and then recover overnight. The best time to water is in the early morning or in the evening when the sun is not intense. Avoid watering during the hottest part of the day.

Fertilizer. Dry habitat plants get very little fertilizer in the wild, and in fact, too much fertilizer can harm them. Furthermore, fertilizer seeps into the ground very rapidly in dry soils and may contaminate ground water. If you wish to fertilize, be sparing. Use only slow-release fertilizers and avoid high-nitrogen fertilizers, which stimulate too much top growth and not enough root development. If fertilization is needed, fertilize in spring, fall, or both, but not in summer or winter.

Mulch. Native plant landscaping in dry areas does not require large quantities of mulch as is used in traditional landscaping. Excess mulch eventually adds moisture-retaining organic material to the soil, making it less suitable for dry-area natives and possibly even causing the roots to rot. A thin layer of pine needles or fallen leaves is often sufficient to discourage weeds and minimize dust. Take care to keep mulch from touching the plant stems, because mulch holds moisture and can cause the stems to rot, especially during periods of high humidity or heavy rainfall.

Pests. Your landscape is part of an ecosystem with each part dependent on every other part, so it is important to think of the effects of any action on the entire system. For example, if you use pesticides near your butterfly garden plants, your butterfly population will decline. Furthermore, pesticides kill beneficial insects and may cause birds to become ill or die from eating poisoned insects. When natural predators such as beneficial insects and birds die, more and more pesticides are needed to control undesirable insects, leading to ever greater expenditures of time and money and the increasing threat of pesticides migrating to ground water.

For the health of your plants, native wildlife, your own family and pets, and our water supply, avoid pesticides whenever possible. If you feel the need to control pests, educate yourself about pest-control methods, both natural and synthetic, or consult with an expert on Integrated Pest Management (IPM). IPM uses multiple methods of insect control, including traps and barriers, beneficial predators, botanical insecticides, sanitation, mechanical controls, and selection of appropriate plants for the site. IPM sometimes uses manufactured insecticides, but only as a last resort. In such cases, IPM favors using the insecticide that will do the job with the least danger to natural pest predators and to the environment and favors limiting insecticide applications to the affected spot. IPM strongly discourages preventive application of insecticides to healthy plants, a practice that wipes out beneficial insect populations. For further information about IPM, consult your county Cooperative Extension Service.

If you avoid unnecessary insecticide applications and learn to tolerate some nibbling on the leaves and fruit, the natural system will eventually reach a balance in which your plants and other creatures can live harmoniously with each other and with you, a major goal of native plant landscaping.

SEASONAL CHANGES

With native plants, you will enjoy new spring leaves, fall color changes, and a seasonal progression of assorted flowers and fruits. The plant descriptions tell which plants go dormant in winter, so you can save their places and watch for them to return.

Trees

American Holly

Common Name: American holly

Botanical Name: *Ilex opaca*

Family: Holly (Aquifoliaceae)

Plant Communities: Maritime forest, mesic hammock

Mature Size (H x W): 40 ft. x 30 ft.

Growth Habit: Tree

Growth Rate: Moderate

Description: American holly is an evergreen, dioecious tree (male and female flowers on different trees) with gray bark and bright green leaves that have sharp points along the margins. The leaves are stiff and arranged alternately along the twigs. Small white flowers produce bright red berries that ripen in the fall/winter on female trees.

Landscape and Wildlife Use. The berries are a food source for birds during the winter, but are poisonous when consumed by humans. They complement southern red cedar and clump grasses, such as seaoats, as wildlife food sources in the landscape. Mockingbirds and robins may be observed defending their berry food source from other birds.

Light: Full sun to partial shade

Soil: Poor, sandy, well drained

Propagation: Seeds, cuttings

Notes: American holly ranges throughout the eastern United States from Virginia south to Florida and west to Texas. In Florida, the southern limit of its natural range is the center of the peninsula, but it can be grown successfully further south. Propagation by seed usually requires stratification (storing seeds at 4° C until spring); the sex of the seedling cannot be determined until flowering—at least 4–5 years. To ensure the production of berries on female trees, plant a male tree along with the female trees.

Yaupon holly *(Ilex vomitoria)*, a related species with a similar range in Florida, has leaves stiff and dark green on the upper surface and dull green on the lower surface. They are up to 1 inch long by 1/2 inch wide with crenate edges. Flowers are small, white. Berries are red and ripen in the fall. This species can be grown as a small tree, or as a thicket-producing shrub. As a shrub, it provides good cover for birds and small mammals. It will also do well in damper sites. An infusion of the leaves is used by Native Americans as an emetic, hence the specific name *vomitoria*. Two notable cultivars are available: weeping yaupon (*Ilex vomitoria* 'Pendula') and dwarf yaupon (*Ilex vomitoria* 'Schellings').

Live Oak

Common Name: Live oak

Botanical Name: *Quercus virginiana*

Family: Oak or beech (Fagaceae)

Plant Communities: Mixed hammocks, pine flatwoods, coastal strand

Mature Size (H x W): 80 ft. x 100 ft.

Growth Habit: Large tree with a rounded crown

Growth Rate: Moderate

Description: Live oak, considered an evergreen, drops its leaves immediately before new spring growth appears. The mature leaves are dark green, leathery, 2 to 5 inches long, 1 to 1 1/2 inches wide, egg-shaped to oblong. The bark is dark and furrowed. Branches spread widely on isolated trees and may even touch the ground.

Landscaping and Wildlife Use: Plant live oak as a specimen or shade tree. Its horizontal branches provide homes for bromeliads, orchids, and ferns and are used as perching places for squirrels and birds, both of which dine on the acorns. People appreciate the live oak for its picturesque shape and copious shade, and children find its low, spreading branches an easy climb. This "grandfather tree" is unsurpassed for large areas.

Light: Full to partial sun

Soil: Well drained to seasonally wet, acid to alkaline

Propagation: From acorns, grafting

Notes: Live oaks are so named because they may live for several hundred years.

Myrtle Oak

Common Name: Myrtle oak

Botanical Name: Quercus myrtifolia

Family: Oak or beech (Fagaceae)

Plant Communities: Sand pine or oak scrubs, coastal strand, sandhills, pine flatwoods, beach dunes

Mature Size (H x W): 40 ft. x 8 ft.

Growth Habit: Large shrub or small tree

Growth Rate: Moderate

Description: This evergreen oak may form shrubby thickets in very dry habitats or grow into a tree in moister soil. Its oval, dark green leaves are 3/4 to 2 inches long and 1/2 to 1 inch wide.

Landscaping and Wildlife Use: Plant as a barrier hedge, a small shade tree, or a substitute for live oak where space is limited. Myrtle oak supplies cover, nesting, and food for birds and mammals, and larval food for several species of hairstreak butterflies, including the white M, the red-banded, the gray, and the Southern oak hairstreaks.

Light: Full to partial sun

Soil: Sandy, well drained, acid to alkaline

Propagation: From acorns

Notes: The acorns are important food for two species listed as threatened by the Florida Game and Fresh Water Fish Commission: the Florida scrub jay and the black bear.

Persimmon

Common Name: Persimmon

Botanical Name: Diospyros virginiana

Family: Ebony (Ebenaceae)

Plant Communities: Maritime forest, flatwoods, sandhill

Mature Size (H x W): 50 ft. x 40 ft.

Growth Habit: Tree

Growth Rate: Moderate

Description: The persimmon's leaves are deciduous and simple, arranged in an alternating pattern along the twigs. They may be up to 6 inches long by 3 inches wide and are lanceolate in shape. The flowers are small, bell-shaped, and greenish yellow. Persimmon bears an edible fruit, 1 1/2 inches in diameter, which is bright orange when ripe. This fruit, which ripens in the fall, is quite sweet when ripe, but is bitter and unpalatable when green.

Landscape and Wildlife Use: The fruit contains several dark-colored seeds, which are widely dispersed by animals that eat the fruit. It is relished by whitetail deer, raccoons, foxes, skunks, many birds, and rodents. Persimmons seem to like an open site that is disturbed periodically. Plant in a wildlife garden or use as an interesting specimen clump. The leaves and fruit will show fall color and the bark on older trees grows in a block pattern.

Light: Full sun to partial shade

Soil: Poor, sandy, well drained

Propagation: Seeds, cuttings

Notes: Persimmon ranges from New England south to Florida and through the Midwestern states. In Florida, it occurs nearly throughout the state. There are only a few problems with persimmon trees. In damp conditions, a sooty mold often grows on the leaves resulting in dark splotches. Tent caterpillars sometimes attack this tree.

Red Bay

Common Name: Red bay

Botanical Name: Persea borbonia

Family: Laurel (Lauraceae)

Plant Communities: Maritime forest, sandhill, scrub

Mature Size (H x W): 50 ft. x 40 ft.

Growth Habit: Tree

Growth Rate: Moderate

Description: Red bay is a medium-sized, evergreen tree. The flowers are greenish yellow, and inconspicuous. The fruit is a round, 1/2-inch-diameter, green drupe that turns black when ripe. It contains a stone holding a single seed. The leaves release an aromatic oil when crushed.

Landscape and Wildlife Use: It grows naturally in dry, sandy sites along the coast and inland xeric sites such as sandhill and scrub. Red bay is relatively easy to establish in the home landscape and requires little care afterwards. The seeds are eaten by squirrels, quail, and other birds. Butterflies, such as the Palamedes, or laurel swallowtail, use red bay as a larval plant. Because red bay is subject to gall, it is best used as a background tree in a woodland garden.

Light: Full sun or partial shade

Soil: Poor, sandy, well drained

Propagation: Cuttings, seeds

Notes: Red bay ranges from North Carolina south throughout Florida and east to Texas. The leaves can be used in cooking, and it is much easier to grow than the traditional bay leaf. A black sooty mold sometimes grows on the leaves, and insects may attack the leaves, resulting in a convoluted growth or gall.

Redbud

Common name: Redbud

Botanical Name: *Cercis canadensis*

Family: Pea (Fabaceae)

Plant Communities: Maritime forest, mesic hammock

Mature Size (H x W): 30 ft. x 20 ft.

Growth Habit: Tree

Growth Rate: Moderate

Description: Redbud is a small deciduous tree. The leaves are roughly heart-shaped with a pointed end, are light to medium green in color, and alternate along the twigs in a zigzag pattern. Redbud is prominent in the spring in our area because of its colorful, 1/2-inch-long, lavender-pink blossoms, which appear before the new leaves unfold. When in full bloom, clusters of flowers along the branches and trunk produce a spectacular display of color. The fruit is a flat pod, 2 to 4 inches long, containing several seeds.

Landscape and Wildlife Use: This tree prefers a rich, moist soil but is adaptable to drier, poorer sites. It is naturally an understory tree but does well as a specimen planting in partial shade. Try it with pine or cedar and coontie.

Light: Full sun to partial shade

Soil: Rich to poor

Propagation: Seeds

Notes: Redbud occurs nearly throughout the eastern United States. In Florida, it performs best in the northern part of our area, but plantings have been successful as far south as the Tampa area.

Red Cedar

Common Name: Red cedar

Botanical Name: Juniperus virginiana (syn. *J. salicicola*)

Family: Cedar (Cupressaceae)

Plant Communities: Maritime forest, coastal strand

Mature Size (H x W): 90 ft. x 25 ft.

Growth Habit: Tree

Growth Rate: Moderate

Description: Red Cedar is an aromatic, evergreen conifer. The bark of the trunk is reddish brown in color and shreds in thin strips. It is typically a canopy tree with a conical- or pyramidal-shaped crown. The lower branches will droop to touch the ground. Leaves are needlelike, pointed, and green. Red cedar is dioecious, having male and female cones on separate trees. Female trees ultimately produce a bluish, berrylike cone in the fall.

Landscape and Wildlife Use: This tree will eventually grow to a very large size, so room for growth should be provided. The fruits are used by cedar waxwings, quail, and turkeys. Many birds also use cedar for cover and nesting. The olive hairstreak butterfly uses red cedar as a larval plant.

Light: Full sun

Soil: Alkaline, well drained

Propagation: Seeds

Notes: Red cedar ranges through the eastern United States into the Great Plains states. It ranges nearly throughout Florida, but is most abundant near the coast, preferring alkaline soils. Red cedar can be used as a potted holiday tree if kept in a bright room for no more than 3 weeks. After the holidays it can be planted in the landscape.

Sand Live Oak

Common Name: Sand live oak

Botanical Name: *Quercus geminata*

Family: Oak or beech (Fagaceae)

Plant Communities: Maritime forest, coastal strand, sandhill, scrub

Mature Size (H x W): 30 ft. x 20 ft.

Growth Habit: Tree

Growth Rate: Moderate

Description: In the coastal uplands, sand live oak grows as a short, shrubby tree, often occurring in thickets. Away from the coast, it can grow to much larger sizes. The leaves are stiff, with rolled edges and veins prominently depressed into the upper surface. Considered a semi-evergreen or evergreen, the leaves usually fall from the tree at the end of winter when new leaves start to develop. Flowers are green and inconspicuous with male and female flowers on the same tree.

Landscape and Wildlife Use: Sand live oak needs no care after being established. It makes an excellent substitute for the larger live oak in small yards. The acorns are an important food source for squirrels and other animals. Several hairstreak butterflies use the leaves as a larval food source.

Light: Full sun

Soil: Poor, sandy, well drained

Propagation: Acorns, cuttings

Notes: Sand live oak ranges from North Carolina south to Florida and west to Louisiana. It is common nearly throughout Florida.

The similar ***live oak (Quercus virginiana)*** is an evergreen tree with a sturdy trunk and stout limbs that grow horizontally, sometimes returning to the ground at their extremities. The leaves fall as new growth is forming at the end of winter. Isolated trees grow short and wide. Live oaks provide lots of shade and are the stately trees characteristic of the south. There are fewer pests that attack live oak, although a moth larva can be a problem in the fall after the acorns appear. It is every child's ideal climbing tree, with a trunk that splits near the ground and strong branches that are widely spaced. Live oak is extremely drought tolerant. It ranges from Virginia south to Florida and west to Texas. It is common nearly throughout Florida.

Slash Pine

Common Name: Slash pine

Botanical Name: Pinus elliottii

Family: Pine (Pinaceae)

Plant Communities: Maritime forest, flatwoods

Mature Size (H x W): 60 ft. x 40 ft.

Growth Habit: Tree

Growth Rate: Moderate

Description: Slash pine is an evergreen conifer. The branches are sturdy, often extending horizontally from the main trunk. Leaves are needles, 2 or 3 per fascicle (bundle), at least 5 inches long. The cones are 3 to 6 inches long.

Landscape and Wildlife Use: Along the coast, this pine occurs in dry uplands, but inland it is usually found on wetter sites. It will adapt to most available landscape uses. The cones and seeds provide an important food source for wild turkeys, gray squirrels, and fox squirrels.

Light: Full sun

Soil: Poor, sandy, well drained

Propagation: Seeds

Notes: Slash pine ranges from North Carolina to Florida and west to Texas. The slash pine native to central and southern Florida is sometimes referred to as South Florida slash pine and is considered by some to be a distinct variety (var. *densa*). It has slightly thicker bark and denser needle growth around the growing tips, which help it survive frequent fires better than slash pines found in northern Florida (var. *elliottii*). The seedlings of the South Florida slash pine are slow growing for several years, remaining in a protective grassy stage similar to longleaf pine, although of much shorter duration.

Southern Magnolia

Common Name: Southern magnolia

Botanical Name: *Magnolia grandiflora*

Family: Magnolia (Magnoliaceae)

Plant Communities: Maritime forest, mesic hammock

Mature Size (H x W): 75 ft. x 35 ft.

Growth Habit: Tree

Growth Rate: Moderate

Description: Southern magnolia is a large evergreen tree. Branches grow low on the trunk, giving the tree a wide conical shape. The large (6 to 9 inches), leathery leaves are dark glossy green on the upper surface and usually with a cinnamon-brown "felt" on the lower surface. The late spring flowers are beautiful, large (to 8 inches across), white, and very fragrant, although some people are allergic to the blooms. Bright red seeds are produced in a conelike fruit.

Landscape and Wildlife Use: Southern magnolia is often used as a single specimen providing a focal point in the landscape. The tree can be trimmed up or allowed to droop to the ground. The bright red seeds are used sparingly by many animals, including turkeys, quail, squirrels, and opossums.

Light: Partial shade (especially when small) to full sun

Soil: Sandy, well drained to rich moist

Southern magnolia continued on next page

Propagation: Seeds, grafting

Notes: Southern magnolia is considered the signature tree of the deep South, ranging from Maryland south to Florida and west to Texas. In Florida, it ranges south to the central peninsula. Cut twigs with leaves and blooms are often displayed in bowls in the home. Pine, palmetto, and wild coffee make attractive companions. There are many good selected cultivars for Florida, including 'Bracken's Brown Beauty,' 'Little Gem' (dwarf), 'Saint Mary,' and 'D.D. Blancher' (tm).

Sugarberry

Common Name: Sugarberry, hackberry

Botanical Name: Celtis laevigata

Family: Hackberry (Celtidaceae)

Plant Communities: Maritime forest, floodplain forest, mesic hammock

Mature Size (H x W): 80 ft. x 40 ft.

Growth Habit: Tree

Growth Rate: Moderate to fast

Description: Sugarberry is a deciduous tree found in many habits ranging from damp to dry. The leaves are arranged in an alternating pattern along the branches. The leaves are lance-shaped, somewhat asymmetrical, and irregularly toothed or smooth. The bark of the young sugarberry is smooth and gray; older branches and trunks have a bark that is broken by irregular corky growths. Sugarberry will flower and bear fruit when young. Flowers are very small, green, and are borne in the axils of the leaves. The fruit is an orange to brown-red drupe about 1/4 to 3/8 inch in diameter.

Landscape and Wildlife Use: The fruit is edible and serves as a food source for mammals and birds such as mockingbirds, robins, thrashers, cardinals, and sapsuckers. The hackberry butterfly uses the leaves as a larval food source. Sugarberry grows fast and can break under stress from high winds. It makes an excellent tree for quick shade and to attract birds.

Light: Full sun

Soil: Sandy, poor to moderate

Propagation: Seeds, cuttings, air layering

Notes: Sugarberry ranges from Maryland south to Florida and west to Texas. It occurs nearly throughout Florida. The wood is used for crates and furniture. Native Americans used the fruit as a sweetener.

Sweet Acacia

Common Name: Sweet acacia

Botanical Name: *Acacia farnesiana*

Family: Pea (Fabaceae)

Plant Communities: Coastal strand, maritime forest

Mature Size (H x W): 15 ft. x 15 ft.

Growth Habit: Small tree

Growth Rate: Slow to moderate

Description: Sweet acacia is a small tree or large shrub with deciduous leaves and prominent spines or thorns on the branches. The leaves are arranged in an alternate pattern along the branches. The leaves are compound, with many pairs of small leaflets. Flowering occurs in spring or sometimes after a drought. The blooms look like bright yellow puff-balls, 1/2 inch in diameter, on a 1/2-inch-long stalk. The fruit develops as a 2- to 3-inch-long cylindrical pod, chocolate-brown in color.

Landscape and Wildlife Use: The seeds of sweet acacia are valuable to wildlife as a food source, and the thorny branches provide good protection for nesting birds and other critters. This tree establishes easily in the home landscape. Weekly watering for a month is enough to get it established. It will bloom in the second year after planting.

Light: Full sun

Soil: Sandy, well drained

Propagation: Seeds, cuttings

Notes: Sweet acacia ranges from southeastern coastal Georgia south throughout Florida and west along the southern tier of states to California. It also occurs in the West Indies, Mexico, Central America, and South America. The fragrant blooms are used in the perfume industry. The durable wood is used for fenceposts and farm tools.

Shrubs

Adam's Needle

Common Name: Adam's needle

Botanical Name: *Yucca filamentosa*

Family: Agave (Agavaceae)

Plant Communities: Scrubs, sandhill, dry, open woods

Mature Size (H x W): 2 ft. x 2 ft.

Growth Habit: Compact shrub

Growth Rate: Moderate

Description: This shrub has stiff, swordlike leaves 12 to 18 inches long with numerous threads along the leaf margins. The single flowering stalk reaches up to 10 feet in height; it is covered with 2-inch-long white flowers on short branches. After blooming, the mother plant slowly dies but is replaced by several suckering offspring.

Landscaping and Wildlife Use: Clump several plants together or combine with evergreen shrubs, flowering shrubs such as garberia, or groundcovers such as coral honeysuckle.

Light: Full sun

Soil: Sandy, well drained

Propagation: From seed, root suckers

Notes: The yucca moth lays eggs within the Adam's needle flower and then pollinates the plant, ensuring that seeds will be available to feed the larvae. Indians used the leaf "threads" for basketry, clothing, ropes, threads, and lines. Adam's needle fiber is considered the strongest in North America. People can eat the young flowers raw in salads or cooked.

Bay Cedar

Common Name: Bay cedar

Botanical Name: *Suriana maritima*

Family: Bay cedar (Surianaceae)

Plant Communities: Beach dune, coastal strand

Mature Size (H x W): 20 ft. x 10 ft.

Growth Habit: Shrub or small tree

Growth Rate: Very slow

Description: Though not a true cedar, this shrub's mass of ascending branches form a dense rounded head that looks like a juniper (cedar), and the crushed leaves have a cedarlike odor. The grayish, narrow, succulent leaves form a tassel-like arrangement at the end of the branches. Even in the southern peninsula, bay cedar loses some of its leaves during the winter months. Small yellow flowers appear throughout the year and are followed by fuzzy, cone-shaped fruits containing tiny black seeds.

Landscape and Wildlife Use: This shrub tolerates harsh, windy conditions from coastal central Florida southward. Use bay cedar as a medium-sized hedge or accent plant in a hot, sandy spot or beachfront. It grows with direct salt spray. Railroad vine, blue porterweed, and dune sunflower are complementary in a natural landscape. The blue and gray hairstreak butterfly uses bay cedar as a larval plant in the southern portion of our area.

Light: Full sun

Soil: Rocky or sandy, well drained, wide range of pH

Propagation: Reputed to be difficult. Seeds collected in July and sprinkled on the surface of sandy soil germinate in approximately three weeks.

Notes: In Florida, bay cedar occurs southward along the coasts from Pinellas and Hillsborough Counties on the west and Indian River County on the east. It also occurs in coastal areas of the West Indies and Mexico. Leaves, bark, and flowers have been used in various medicinal preparations to treat intestinal disorders, fevers, toothache, and other conditions. The wood is carved into hooks and small figures.

Beautyberry

Common Name: Beautyberry

Botanical Name: *Callicarpa americana*

Family: Mint (Lamiaceae)

Plant Communities: Coastal strand, maritime forest, flatwoods

Mature Size (H x W): 7 ft. x 6 ft.

Growth Habit: Bushy to open shrub

Growth Rate: Fast

Description: A coarse-leaved, deciduous shrub, beautyberry's toothed, ovate (egg-shaped) leaves grow opposite one another on arching stems. Fine, white hairs cover young stems, leaves, and flower parts. The small, lavender-pink flowers grow close to the stem in clusters during late spring. These are followed by conspicuous purple, magenta, or, occasionally, white berries that persist on leafless winter stems until eaten by birds.

Landscape and Wildlife Use: Plant beautyberry in full sun with seashore goldenrod, saw palmetto, and dwarf Fakahatcheegrass or grouped in a woodland planting with wild coffee and wild plumbago. This versatile shrub grows almost anywhere and provides not only winter color, but a food source for birds after everything else has been eaten. Robins, mockingbirds, catbirds, brown thrashers, bobwhite quail, and eastern towhees occasionally consume the berries. Beautyberry can be pruned before new leaves emerge to encourage a bushy, compact form. Plant in full sun for heaviest fruiting.

Light: Full sun to full shade

Soil: Sandy to partially organic, acid to alkaline; tolerates moisture but not flooding

Propagation: Clean flesh from ripe seed and plant in spring or wait for birds to shed seeds and harvest seedlings in spring; easy to grow from softwood cuttings.

Notes: It occurs from Maryland south throughout Florida and west to Texas. It is the only species of *Callicarpa* native to the United States. Bees are attracted to the abundant blooms. A tasty jelly can be made from beautyberry fruit.

Christmasberry

Common Name: Christmasberry

Botanical Name: *Lycium carolinianum*

Family: Nightshade (Solanaceae)

Plant Communities: Beach dunes, coastal strand, salt flats

Mature Size (H x W): 9 ft. x 3 ft.

Growth Habit: Upright, sprawling shrub

Growth Rate: Slow

Description: Christmasberry is a spiny shrub with small, fleshy, club-shaped leaves attached directly to the stem in an alternating arrangement. The leaves are not shed until late winter, if at all. When crushed, the leaves give off a pungent smell. Blue, lavender, or white tubular flowers appear in the fall and are followed by bright red, 1/2-inch-long egg-shaped berries.

Landscape and Wildlife Use: Plant in groups surrounded by clumping and spreading grasses. For interesting texture and color contrast, try adding corkystem passionflower, blanketflower, or other seasonal flowering natives. The Canada goose is reported to eat the berries along the Gulf Coast.

Light: Full sun to filtered shade

Soil: Sandy or shelly, well drained to periodically wet

Propagation: Seeds, cuttings

Notes: Christmasberry grows in coastal areas from South Carolina south throughout Florida and west to Texas. It also occurs in the West Indies. This member of the nightshade family may be toxic.

Coontie

Common Name: Coontie

Botanical Name: *Zamia pumila*

Family: Zamia (Zamiaceae)

Plant Communities: Hammocks, pine flatwoods, Indian middens

Mature Size (H x W): 1 1/2 ft. x 3 ft.

Growth Habit: Mounding shrub

Growth Rate: Slow

Description: Coontie is a low, evergreen, fernlike shrub with leathery leaves. New leaves emerge from the center of the plant in growth spurts once or twice a year. Sometimes they are covered with rusty fuzz that is shed as they expand, becoming a dark, glossy green. Male and female "flowers" are produced in conelike structures on separate plants. The female cones produce large red seeds that are poisonous to people if eaten.

Landscaping and Wildlife Use: Coontie is popular for its tropical look. It is perfect as a specimen plant, in mass plantings, in drifts with other groundcovers, around pines or tall palms, or in medians.

Light: Full to partial sun

Soil: Sandy, well drained, acid to alkaline

Propagation: Very slow from seed. Allow fruit to wither, then remove skin and fleshy pulp from seed before planting. An alternative method is to dry fruit in a basket or mesh bag for several months, crack with a nutcracker, then plant.

Notes: American Indians harvested the coontie's large, woody, underground stem, removed the poisonous compounds, and pounded it into flour. Settlers in south Florida manufactured starch from the stem. Coontie belongs to one of the oldest plant families on earth, the cycads, sometimes called "dinosaurs of the plant kingdom." In south Florida, coontie leaves provide larval food for the atala butterfly.

Coralbean

Common Name: Coralbean or Cherokee bean

Botanical Name: *Erythrina herbacea*

Family: Pea (Fabaceae)

Plant Communities: Coastal strand, dry to mesic (i.e., moderately moist) hammocks, pine flatwoods

Mature Size (H x W): 16 ft. x 4 ft. (unless pruned back by frost)

Growth Habit: An upright shrub with an open growth habit

Growth Rate: Fast

Description: The most distinctive feature of this deciduous shrub is a spring display of showy red flower spikes. The brown seed pods open in summer to reveal bright scarlet seeds. The branches and stems may have spines. Each compound leaf has three arrowhead-shaped leaflets. Coralbean dies back to the ground after extreme cold but returns from the roots in spring.

Landscaping and Wildlife Use: Plant in front of large shrubs or in coastal zones or to attract hummingbirds. Combine with pink muhlygrass, sand cordgrass, blue porterweed, or tropical sage. Prune to any desired height. The large root system and long taproot make transplanting difficult.

Light: Full to partial sun

Soil: Sandy, well drained, acid to alkaline

Propagation: From seed, stem cuttings. Nick each seed with a file (i.e., scarify it) or douse it with boiling water before planting. If kept moist, it will germinate within seven days.

Notes: The seeds are poisonous.

Firebush

Common Name: Firebush

Botanical Name: *Hamelia patens*

Family: Madder (Rubiaceae)

Plant Communities: Maritime forest, coastal strand

Mature Size (H x W): 12 ft. x 8 ft.

Growth Habit: Large, bushy shrub

Growth Rate: Fast

Description: Reddish stems and leaf veins, set off by scarlet to orange tubular flowers, give this showy shrub its name. Firebush's 3- to 6-inch-long leaves, which are also sometimes reddish colored, grow in whorls of 3 to 7 around the stem. The red-orange flowers are followed by 1/4-inch dark purple fruits.

Landscape and Wildlife Use: Firebush looks best in filtered shade, but will tolerate a wide range of light, soil, and moisture conditions. Use it in wildlife plantings with slash pine, sweet acacia, and wild plumbago, or as a flowering hedge in warmer areas. Freezing temperatures will kill firebush to the ground, but established plants grow back even bushier in late spring. Firebush is an excellent foundation shrub next to the house or along a fence. If you don't usually receive winter freezes, prune during spring or summer to control the size. This shrub blooms only on new wood and tends to get leggy if not maintained. The bright flowers are favored by both hummingbirds and butterflies, while birds enjoy the fruit. Firebush attracts multitudes of zebra longwing butterflies in shady locations. If you have room for only one shrub in your butterfly garden, this is the one.

Light: Full sun to shade

Soil: Sandy to organic, acid to alkaline

Propagation: Cuttings, seeds; often self-seeds or is spread by birds

Notes: It occurs on coastal sites in central and southern peninsular Florida, rarely growing inland. It also occurs in Mexico, Central America, and South America.

"African firebush" or "dwarf firebush" (H. patens var. glabra), is a nonnative similar-appearing plant from Central and South America. It is smaller with glossy leaves and yellowish flowers.

Florida Privet

Common Name: Florida privet, Florida swampprivet

Botanical Name: *Forestiera segregata*

Family: Olive (Oleaceae)

Plant Communities: Beach dune, coastal strand, maritime forest, rocky pineland

Mature Size (H x W): 20 ft. x 10 ft.

Growth Habit: Large shrub

Growth Rate: Moderate

Description: Considered semi-evergreen, Florida privet sometimes loses its leaves during the winter flowering period. Tiny, greenish yellow flowers occur in small clusters along the stem. After flowering, shiny, 1- to 3-inch-long elliptical leaves grow opposite one another on the stem. The fruits resemble a small, 1/4- to 1/2-inch-long, blue-black olive. Male and female flowers occur on separate plants; consequently, fruits are only found on female plants.

Landscape and Wildlife Use: Florida privet can be left in its natural form or sheared for use in hedge and foundation plantings. It is excellent in wildlife plantings, where it provides nesting and cover habitat, as well as a nectar and fruit source for insects, birds, and small mammals. Ducks, quail, robins, and deer occasionally eat the fruits. Red hairstreak butterflies favor the flowers and serve as a pollinator. Combine Florida swamp privet with native grasses and flowering plants such as sand cordgrass, seaoats, firebush, and pricklypear cactus.

Light: Full sun to shade

Soil: Sandy to organic, acid to alkaline; tolerates periodic wet conditions after establishment

Propagation: Remove seed from fleshy fruit and sow immediately.

Notes: Florida privet is common from South Carolina south throughout coastal Florida. It tolerates harsh beachfront conditions, periodic inundation, and drought.

Inkberry

Common Name: Inkberry, beachberry

Botanical Name: *Scaevola plumieri*

Family: Goodenia (Goodeniaceae)

Plant Communities: Beach dune, coastal strand

Mature Size (H x W): 6 ft. x 6 ft.

Growth Habit: Stout, succulent shrub

Growth Rate: Slow to moderate

Description: This evergreen has alternate, obovate, glossy green leaves to 6 inches long clustered near the branch tips. The interesting white to pinkish, tubular flowers, about 1 inch long in the axils of the leaves, are split down one side and fan-shaped. The fruit is a black, egg-shaped, 2-seeded drupe about 1/2 inch long.

Landscape and Wildlife Use: Although considered an excellent dune stabilizer, inkberry is adaptable to other sandy, full-sun areas, where it will serve as a groundcover or low-growing shrub. Birds and small mammals eat the juicy black fruits.

Light: Full sun

Soil: Sandy, well drained, wide range of pH

Propagation: Seed germinates in about 6 weeks. Rooted stems can be detached from the mother plant, or cuttings can be rooted in sandy soil.

Notes: Inkberry ranges along the Florida coast from Pinellas and Brevard Counties southward. It also occurs in the West Indies, Mexico, Central America, South America, Africa, and Sri Lanka. This is the only representative of the genus native to Florida; most are from Australia. It differs from the invasive exotic species, beach naupaka (*S. taccada*), by producing black fruits, whereas the pesty beach naupaka bears white fruits.

Marlberry

Common Name: Marlberry

Botanical Name: *Ardisia escallonioides*

Family: Myrsine (Myrsinaceae)

Plant Communities: Maritime forest

Mature Size (H x W): 15 ft. x 5 ft.

Growth Habit: Large shrub to small tree

Growth Rate: Moderate

Description: Marlberry is an evergreen shrub with leathery, 4- to 6-inch, shiny leaves topped by fragrant white to pink flowers at any time during the year. The flowers are followed by clusters of small red to dark purple fruit.

Landscape and Wildlife Use: Used as a specimen or hedge plant in a shady location, marlberry is unsurpassed. Birds and small mammals eat the dry, acid fruits. Although the stems are not sturdy enough for nesting, it does provide good cover. Marlberry forms a small, columnar tree in south Florida.

Light: Partial sun to full shade

Soil: Sandy, alkaline

Propagation: Seeds; slow to germinate

Notes: In Florida, Marlberry occurs from Flagler County on the east coast and Pasco County on the west coast southward. It also occurs in the West Indies and Mexico. Miccosukee Indians added the leaves of this plant to their tobacco as an extender.

Necklacepod

Common Name: Necklacepod

Botanical Name: *Sophora tomentosa* var. *truncata*

Family: Pea (Fabaceae)

Plant Communities: Beach dune, coastal strand

Mature Size (H x W): 10 ft. x 6 ft.

Growth Habit: Upright shrub

Growth Rate: Slow

Description: Necklacepod is an evergreen shrub with glossy, dark green leaves that alternate on the stem and are segmented into 11 to 21 leaflets. Young leaves are covered with grayish hairs—hence the species name *tomentosa*. Clusters of yellow, pealike flowers 4 to 8 inches long grow from the tips of the branches most of the year, followed by strings of brown seed pods that resemble necklaces.

Landscape and Wildlife Use: This under-used shrub is an excellent hedge or specimen plant in a sunny, open area; it will not survive if planted in shade. Surround it with low-growing groundcover plants such as saltgrass or knotgrass and railroad vine or blue porterweed. Necklacepod is becoming rare in some natural areas because of crowding by nonnative, invasive plants like Brazilian pepper. A variety of wildlife uses the leaves and flowers, including hummingbirds, which have been observed fighting over the flowers. Necklacepod is frost sensitive and may freeze to the ground but will return quickly from the base the following spring.

Light: Full sun

Soil: Sandy, alkaline

Propagation: Easy from seed sown in pots or directly into sandy soil

Notes: Necklacepod is a shrub of warm coastal areas in central and southern Florida and the West Indies. *Sophora* is an Arabic word referring to part of the flower, which resembles a butterfly. The seeds are poisonous to humans. The pods are interesting in dried flower arrangements.

A related plant, ***Sophora tomentosa* var. *occidentalis***, native only to Texas, is often sold through nurseries. It differs from our native variety in having smaller, permanently silky, pubescent leaves.

Seagrape

Common Name: Seagrape

Botanical Name: *Coccoloba uvifera*

Family: Buckwheat (Polygonaceae)

Plant Communities: Beach dune, coastal strand, maritime forest

Mature Size (H x W): 50 ft. x 20 ft.

Growth Habit: Large shrub or small tree

Growth Rate: Moderate

Description: Seagrape is well known for its thick, evergreen, almost round leaves that are bronze when young, but turn green with red veins as they mature. Yellow-green or off-white flowers hang in 4- to 8-inch-long bunches in the spring, followed by purple 3/4-inch-long fruit, which are edible. They ripen sporadically, one "grape" at a time.

Landscape and Wildlife Use: Use seagrape as a specimen tree or a hedge or a windbreak. It provides cover and food to mammals and birds, while the flowers produce nectar used by honeybees. It is cold sensitive and, at or near the northern end of its natural range, should be placed in protected areas.

Light: Full sun to filtered shade

Soil: Sandy to organic-rich, wide range of pH

Propagation: Seeds, ground layering

Notes: Seagrape occurs from Volusia County on the east coast and Pasco County on the west coast southward. It is also found in coastal areas of the West Indies, Central America, and South America. Seagrape fruits are used in jelly and wine, the hard wood is used for carving and cabinetry, and the sap was originally used in tanning and dyeing. The fruits have been used medicinally to reduce fever, the roots to treat diarrhea, and the bark in an extract to treat dysentery.

Simpson's Stopper

Common Name: Simpson's stopper, twinberry

Botanical Name: *Myrcianthes fragrans*

Family: Myrtle (Myrtaceae)

Plant Communities: Maritime forest

Mature Size (H x W): 20 ft. x 6 ft.

Growth Habit: Large shrub or small tree

Growth Rate: Moderate

Description: Fragrant flowers, interesting bark, and showy red fruit set off Simpson's stopper. Reaching tree size in hammock shade, it grows smaller in the sun, where its 1- to 3-inch-long glossy, evergreen leaves take on a coppery color. Masses of 1/2-inch white flowers cover the plant in spring, followed by small, bright orange-red fruit. Like many other members of the myrtle family, Simpson's stopper has flaky bark.

Landscape and Wildlife Use: Grow this stopper as a sheared hedge in a sunny landscape, or a large, open shrub in the shade. Simpson's stopper works well for softening corners of houses or for difficult locations like right-of-ways. As with many native shrubs, this one is a valuable addition to the wildlife garden, where the fruit is promptly eaten by birds. Combine with snowberry, bluecurls, and wild coffee in filtered shade. Larger specimens show off the interesting bark. Pink muhlygrass and coontie are good companions in the sunny garden.

Light: Full sun to full shade

Soil: Sandy to slightly organic

Propagation: Seeds, cuttings

Notes: Rarely found inland, Simpson's stopper grows in coastal hammocks southward from St. Johns County on the east coast and from Charlotte County on the west coast. It also occurs in the West Indies. The **white stopper *(Eugenia axillaris)***, named for its white bark, is common in coastal hammocks in central and southern Florida. Its fruit is edible, and the "skunky"-smelling leaves were once used for the treatment of diarrhea, hence the common name "stopper."

Snowberry

Common Name: Snowberry

Botanical Name: *Chiococca alba*

Family: Madder (Rubiaceae)

Plant Communities: Maritime forest, rocky pineland

Mature Size (H x W): 10 ft. x 10 ft.

Growth Habit: Sprawling shrub

Growth Rate: Slow to moderate

Description: Bright evergreen, 2- to 4-inch leaves grow opposite one another on weak, vinelike stems that often seek support on adjacent plants. Clusters of small, yellow, bell-shaped flowers adorn snowberry most of the year, especially in late spring and summer. These fragrant flowers are followed by 1/4- to 1/2-inch pure white, fleshy fruits.

Landscape and Wildlife Use: Snowberry can be used as a striking groundcover around large trees, such as slash pine and southern magnolia, or scattered with less frequency among native landscapes. Although not particularly favored as a food plant for birds, snowberry provides some cover and interest in the landscape.

Light: Full sun to shade

Soil: Slightly acid to alkaline

Propagation: Slow from seed sown in late winter

Notes: Snowberry is frequently found in coastal areas of central and southern Florida, from Dixie County on the west coast and Duval County on the east coast. It also occurs in the West Indies.

Sparkleberry

Common Name: Sparkleberry, farkleberry

Botanical Name: *Vaccinium arboreum*

Family: Heath (Ericaceae)

Plant Communities: Maritime forest, sandhill

Mature Size (H x W): 30 ft. x 15 ft.

Growth Habit: Large shrub or small tree

Growth Rate: Slow

Description: Sparkleberry's oval, glossy green, 1- to 3-inch-long leaves turn red and fall off in the colder areas of central Florida. As the new leaves emerge, delicate 1/4- to 1/2-inch white, bell-shaped flowers emerge on short sprays in April or May. These are followed by 1/4-inch seedy black fruits in the fall. Sparkleberry has interesting twisted trunks, with reddish flaky bark that is often covered with lichens when growing in the wild.

Landscape and Wildlife Use: Sparkleberry is ideally used at the edge of the woods or as a background plant for flowering plants such as bluecurls, spotted beebalm, blanketflower, and other flowering plants. The flowers are attractive to butterflies, bees, and other nectar-feeding insects, while the bland fruits are relished by birds. This long-lived shrub is interesting in every season, from spring flowers to fall fruits to winter color, with fruits and exposed trunks. Sparkleberry and other *Vaccinium* species are a food source for turkeys, bluebirds, catbirds, fish crows, flickers, flycatchers, kingbirds, jays, robins, pinewoods sparrows, and tufted titmice. Rabbits, opossums, and gray foxes also are known to use *Vaccinium* species.

Light: Full sun to filtered shade

Soil: Acid to alkaline

Propagation: Seeds

Notes: This small tree is found from Virginia south to central Florida and west to Kansas and Texas.

Tough Bully

Common Name: Tough bully

Botanical Name: *Sideroxylon tenax* (syn. *Bumelia tenax*)

Family: Sapodilla (Sapotaceae)

Plant Communities: Maritime forest, coastal strand, sandhill, scrub

Mature Size (H x W): 20 ft. x 15 ft.

Growth Habit: Large shrub or small tree

Growth Rate: Slow

Description: Silky hairs, ranging from tan to copper colored, cover new growth and leaf undersides of tough bully. The evergreen, dark green leaves grow on thorny, crooked branches, with reddish brown, scaly bark. Tiny white flowers emerge in late spring and are followed by 1/2-inch-long black fruit in the fall.

Landscape and Wildlife Use: Tough bully is useful in wildlife plantings, where its stiff branches provide excellent cover and its fruit is eaten by birds. Several plants can be used as a privacy screen or as specimen plants with railroad vine, pricklypear cactus, and Fakahatcheegrass. This shrub is striking when breezes reveal the contrast between deep green upper leaf surfaces and the coppery undersides.

Light: Full sun to partial shade

Soil: Sandy, well drained

Propagation: Seeds

Notes: This species occurs from South Carolina south to nearly throughout peninsular Florida. Although sometimes naturally occurring in maritime forests, it does best when planted in full sun.

Varnishleaf

Common Name: Varnishleaf, Florida hopbush

Botanical Name: *Dodonaea viscosa*

Family: Soapberry (Sapindaceae)

Plant Communities: Beach dune, coastal strand

Mature Size (H x W): 15 ft. x 6 ft.

Growth Habit: Shrub or small tree

Growth Rate: Slow to moderate

Description: The evergreen leaves of varnishleaf are narrow-elliptic or spatulate, revolute, shiny dark green, and are 1 to 6 inches long with a rounded tip. A sticky texture to the leaves gives them the look and feel of varnish. Small yellowish green flowers produce a papery capsule with 3 or 4 wings that starts out yellow, then turns red and finally brown.

Landscape and Wildlife Use: Used as a hedge or specimen plant, varnishleaf is very tolerant of wind and salt spray. The papery seed capsule provides an interesting contrast to the dark green leaves.

Light: Full sun

Soil: Sandy, well drained

Propagation: Collect after seed pod has dried; germination takes about 3 weeks.

Notes: Varnishleaf occurs along both coasts of central and southern Florida from Hernando and St. Johns Counties southward. It is fairly widespread in tropical America. The fruits have been used as a fish poison and medicinally.

Wild Coffee

Common Name: Wild coffee

Botanical Name: *Psychotria nervosa*

Family: Madder (Rubiaceae)

Plant Communities: Maritime forest, hammock

Mature Size (H x W): 10 ft. X 3 ft.

Growth Habit: Shrub

Growth Rate: Moderate

Description: Wild coffee is a variable, evergreen shrub that ranges from a short groundcover to a large shrub in warmer areas. The 3- to 7-inch-long dark green leaves have a quilted appearance similar to gardenia and grow opposite one another, a common characteristic of the coffee family. Tiny white, tubular flowers produce 1/2-inch-long red fruits during summer and fall.

Landscape and Wildlife Use: Plant wild coffee in partial to full shade of live oak or cabbage palm, with wild plumbago, snowberry and bluecurls. It is ideal for a quiet corner, where the zebra longwing butterfly can feed on the nectar and birds can browse on the showy fruits. This plant may freeze to the ground only to return from the roots in the spring, thereby retaining its short stature in central Florida. Hard freezes may kill tender young plants. Use wild coffee as a substitute for the invasive exotic scratchthroat (*Ardisia crenata*).

Light: Filtered sun to full shade

Soil: Acid to alkaline

Propagation: Easy from seed

Notes: Wild coffee ranges throughout peninsular Florida and south through the West Indies, Mexico, and Central America to South America. The wild coffee fruits contain two seeds that resemble commercial coffee beans; however they are too bitter for consumption.

Shortleaf wild coffee ***(Psychotria sulzneri)*** of central and southern Florida and the West Indies produces gray-green velvety leaves and orange-red fruits.

Groundcovers, Vines, and Wildflowers

Beach Verbena

Common Name: Beach verbena, coastal mock vervain, seaside verbena

Botanical Name: *Glandularia maritima* (syn. *Verbena maritima*)

Family: Vervain (Verbenaceae)

Plant Communities: Beach dune, coastal strand, maritime forest

Mature Size (H x W): 1 ft. x 3 ft.

Growth Habit: Perennial creeping groundcover

Growth Rate: Fast

Description: Creeping, square-stemmed groundcover with wedge-shaped, lobed, slightly succulent leaves and clustered rose or purple flowers. This plant blooms almost year-round.

Landscaping and Wildlife Use: Good groundcover for well-drained slopes; can be trimmed back severely. Its flat-topped flower clusters are attractive to butterflies.

Light: Full sun

Soil: Sandy, alkaline, well drained

Propagation: Seeds, creeping stem cuttings

Notes: Beach verbena occurs primarily along the east coast of peninsular Florida. It is similar in description to **Tampa mock vervain *(Glandularia tampensis)*** of the central peninsula, but Tampa mock vervain does not tolerate salt or drought and has a more upright growth habit.

Blanketflower

Common Name: Blanketflower, firewheel

Botanical Name: *Gaillardia pulchella*

Family: aster (Asteraceae)

Plant Communities: Beach dune, coastal strand

Mature Size (H x W): 1 ft. x 2 ft.

Growth Habit: Clumping annual or perennial

Growth Rate: Moderate

Description: Upright, branching, hairy stems bear toothed, narrow, alternating leaves about 3 inches long. The daisylike flowers have yellow to reddish purple straplike ray flowers and crimson central tubular disc flowers. Rarely, all yellow forms occur.

Landscaping and Wildlife Use: This plant may become ragged in appearance after blooming in the fall. It may be sheared back after blooming ends, but wait until seeds are brown and dry to encourage reseeding. It is drought tolerant after establishment and a good butterfly nectar plant.

Light: Full sun

Soil: Sandy, alkaline to acid, well drained

Propagation: Transplant divided clumps or plant collected seeds in spring. It is frequently available in the nursery trade as plants or seeds.

Notes: Blanketflower ranges throughout much of eastern, midwestern, and southern North America and in northern Mexico.

Bluecurls

Common Name: Bluecurls, forked bluecurls

Botanical Name: *Trichostema dichotomum*

Family: Mint (Lamiaceae)

Plant Communities: Maritime forest, sandhill, flatwoods

Mature Size (H x W): 3 ft. x 3 ft.

Growth Habit: Annual with a bushy, rounded habit

Growth Rate: Fast

Description: Bluecurls has square stems with small opposite leaves that are aromatic (like pennyroyal). The numerous lipped and hooded blue flowers have long curled stamens, giving this plant its common name. It blooms in late spring with heaviest blooms in late summer or early fall.

Landscaping and Wildlife Use: This annual makes a good temporary summer hedge; also attractive in front of evergreens. It can be trimmed to shape. If the old plant is left in place, many seedlings will volunteer in spring and summer. It may not tolerate direct salt spray.

Light: Full sun produces heavier blooming plants.

Soil: Sandy, well drained

Propagation: Reseeds freely; collect the seeds after the plant drops its leaves and the dry seed capsules are apparent. Pull up the plant and store upside down in a paper bag for a few days before shaking the seeds loose. The numerous young seedling volunteers appearing in early summer can be transplanted. Seeds and seedlings are slow-growing. Two-inch stem cuttings may also be taken from young plants.

Notes: Bluecurls grows nearly throughout the eastern United States and west to Texas. The small leaves of this plant may be used in potpourri as they are similar in fragrance to pennyroyal.

Blue Porterweed

Common Name: Blue porterweed

Botanical Name: *Stachytarpheta jamaicensis*

Family: Vervain (Verbenaceae)

Plant Communities: Coastal strand, maritime forest

Mature Size (H x W): 2 ft. x 3 ft.

Growth Habit: Sprawling or decumbent, often forming a dense mound

Growth Rate: Fast

Description: The leaves of blue porterweed are light green or gray-green, sometimes blushed with a tinge of purple. They are generally glabrous and oblong or ovate-lanceolate, 1 to 4 inches long, with the margins coarsely serrate. Every day, a few small blue flowers bloom in a whorled arrangement, starting at the bottom of each flowering, quill-like, 5- to 10-inch-long spike.

Landscape and Wildlife Use: Use this tender perennial as a low-growing groundcover, filler plant, or butterfly garden plant. It should be trimmed several times a year to keep the plant at its flowering best and within bounds. Blue porterweed blooms all year unless knocked back by freezing weather, from which it recovers slowly. This plant is a staple butterfly garden nectar plant.

Light: Full sun to partial shade

Soil: Sandy to slightly organic; moist with periodic drying is best once established; does not tolerate prolonged inundation

Propagation: Self-seeding, cuttings, air layering of rooted decumbent stems

Notes: Blue porterweed ranges from central Florida south to the West Indies. This Florida native is often confused with several exotic species that are sold as native. The exotic species are erect shrubs to 6 ft., with glossy, dark green leaves, and flowers varying from deep blue to reddish.

Corkystem Passionflower

Common Name: Corkystem passionflower

Botanical Name: Passiflora suberosa

Family: Passionflower (Passifloraceae)

Plant Communities: Maritime forest

Mature Size (H x W): Vines to 20 feet long, depending on support

Growth Habit: Perennial climbing or trailing, suckering vine

Growth Rate: Moderate

Description: Climbing vine with smooth or sparsely hairy stems. The older stems develop corky wings. Leaves are variable from 1 1/2 to 4 inches in length, smooth-margined, unlobed to 3-lobed. The flowers are small and greenish yellow. The fruits are berries that ripen to purple or black. Corkystem passionflower flowers and fruits most of the year.

Landscaping and Wildlife Use: Plant along a fence or allow to climb into a tree. It tolerates shade and may also be used as an aggressive groundcover. A must for the butterfly garden; caterpillars generally keep the vine "pruned." Zebra longwing butterflies lay their tiny eggs only on passionflowers, and their caterpillar stage feeds on the foliage. The plants may try to "defend" themselves with poisonous foliage or fake eggs on the stems that fool the female into looking elsewhere for a suitable site to lay her eggs. The plants also attract Julia and gulf fritillary butterflies.

Light: Full to partial sun

Soil: Sandy, well drained; tolerant of moist soils

Propagation: Cuttings, seeds

Notes: Corkystem passionflower ranges nearly throughout peninsular Florida and also occurs in Texas, the West Indies, Mexico, Central America, and South America. The edible fruit has been used historically to make ink. The vine suckers and is frequently seeded by birds feeding on the fruit.

Dayflower

Common Name: Dayflower, whitemouth dayflower

Botanical Name: Commelina erecta

Family: Spiderwort (Commelinaceae)

Plant Communities: Beach dune, coastal strand, sandhill, flatwoods, scrub

Mature Size (H x W): 2 ft. x 1 ft.

Growth Habit: Perennial wildflower

Growth Rate: Moderate

Description: The flower stem emerges from a clump of grasslike leaves. The 3-petaled flower has 2 blue petals and a smaller third white petal held in a boat-shaped spathe around the buds and fruit. Each flower blooms in the morning before the hot sun and lasts only one day (hence the common name, dayflower). The plant blooms from early spring to fall.

Landscaping and Wildlife Use: It requires no maintenance and is excellent for naturalizing or in a front border. Songbirds, doves, and quail eat the seeds.

Light: Full sun

Soil: Sandy, well drained

Propagation: Seeds

Notes: Dayflower is common throughout Florida and the eastern United States, west to Arizona. Young tender leaves may be used in salads. Linnaeus named the plant for three Dutch botanist brothers, two of whom were active and productive botanically (two large blue petals) while the third brother never accomplished anything (small white petal).

Dune Sunflower

Common Name: Dune sunflower, beach sunflower

Botanical Name: Helianthus debilis

Family: Aster (Asteraceae)

Plant Communities: Beach dune, coastal strand

Mature Size (H x W): 3 ft. x 6 ft.

Growth Habit: Begins growth as an upright, shrubby annual or perennial, then continues growth as a coarse, rambling stem spreading by aboveground runners that root at the nodes, forming low mats. Grows as an annual in the northern limits of its range.

Growth Rate: Fast

Description: This true sunflower's low rambling stems are often rough and hairy. The 2- to 4-inch rough leaves grow in an alternate pattern on the stem and are triangular in shape with a toothed margin. The 2-inch-wide flower head is composed of 10 to 20 yellow straplike ray flowers and brown to purplish tubular disc flowers in the center. The plant blooms from spring to fall in full sun.

Landscaping and Wildlife Use: Plant as a rampant groundcover where quick cover is desired. It will out-compete slower-growing plants and take over an area via runners and self-seeding. It can be occasionally cut back or clipped to maintain space to encourage denser vegetative leafy growth as old plants become leggy; or remove old plants in the fall/winter and allow new seedlings to germinate in the spring. Dune sunflower is a food source for birds and small mammals and a nectar source for butterflies.

Dune sunflower continued on next page

Light: Full sun

Soil: Sandy, alkaline, well drained

Propagation: Easy from collected seed or cuttings of the runner. Plant March through November due to winter dormancy. Readily self-seeds in the spring; transplant young seedlings.

Notes: There are several subspecies and varieties of dune sunflower differing in stem hairiness, leaf shape, and flower size. The more northerly ***cucumberleaf dune sunflower (subsp. cucumerifolius)*** has a more upright growth habit while the ***west coast dune sunflower (subsp. vestitus)*** and the ***east coast dune sunflower (subsp. debilis)*** are more low growing. The leaves of the west coast subspecies are coarser, more irregularly toothed than those of east coast subspecies. Subspecies *cucumerifolius* ranges from Maine and around the Great Lakes region south to central Florida and west to Texas, while the other two subspecies are restricted to Florida. Subspecies *debilis* occurs along the east coast of Florida to the Keys, while subspecies *vestitus* is found only in the central peninsula on the west coast. *Helianthus* means "sunflower" and *debilis* means "weak" in Latin, describing the low-lying stems. It provides some stabilization of dunes. Unlike the larger-flowering cultivated sunflower, this plant has apparently not been used by people for food or medicine.

Dwarf Fakahatcheegrass

Common Name: Dwarf Fakahatcheegrass, Florida mock gamagrass

Botanical Name: *Tripsacum floridanum*

Family: Grass (Poaceae)

Plant Communities: Rocky pineland

Mature Size (H x W): 5 ft. x 4 ft. (including seedheads)

Growth Habit: Clumping grass

Growth Rate: Fast

Description: Dwarf Fakahatcheegrass is an evergreen, clumping grass with one to many flowering stalks. The leaf blades are 1/4 to 1/2 inch wide and glossy. The green flowering spikes produce a fruit structure with bony, beadlike fruits stacked on one another.

Landscape and Wildlife Use: Dwarf Fakahatcheegrass makes an excellent groundcover in mass plantings with shorter plants such as blue porterweed or dune sunflower planted in front and with spotted beebalm and red salvia interspersed. It also performs well in median strips and other problem areas. Prolonged drought or cold weather will cause dwarf Fakahatcheegrass to turn brown.

Light: Full sun

Soil: Poor, sandy, wide range of pH, dry to wet sites

Propagation: Seeds; germinates in about 2 weeks

Notes: Although native only to rocky pinelands of Collier, Miami-Dade, and Monroe Counties in Florida, dwarf Fakahatcheegrass can be used in the landscape in a wide range of habitats much farther north to central Florida, even in coastal sites. It is very efficient at removing pollutants from water flowing slowly through clumps of the grass. It is listed as a threatened species in Florida.

Marshhay Cordgrass

Common Name: Marshhay cordgrass, saltmeadow cordgrass

Botanical Name: *Spartina patens*

Family: Grass (Poaceae)

Plant Communities: Beach dune, salt flat

Mature Size (H): 3 ft.

Growth Habit: Thin to dense runner, solitary, or in small clumps

Growth Rate: Fast

Description: This evergreen grass usually spreads by thin, wiry rhizomes (underground runners), but it sometimes forms small clumps. The leaves are mostly about 12 to 14 inches long, 1/8 inch wide. The blades are usually flat at the base and become involute (rolled inward from the edges) toward the tip. The green flowering spikes are two to several, somewhat spreading from the main axis of the inflorescence.

Landscape and Wildlife Use: Marshhay cordgrass is well suited for soil stabilization and shoreline protection plantings. It responds well to occasional clipping. Plant from March to November.

Light: Full sun

Soil: Sandy, wide range of pH; tolerates temporary flooding

Propagation: Division

Notes: Marshhay cordgrass occurs from southeastern Canada south to Florida and west to Texas. It also occurs in the Great Lakes area and along the west coast of North America. It was once cut abundantly for hay. It is found exclusively in saline habitats, unlike ***sand cordgrass (S. bakeri),*** which can in found in freshwater wetlands.

Partridge Pea

Common Name: Partridge pea

Botanical Name: *Chamaecrista fasciculata*

Family: Pea (Fabaceae)

Plant Communities: Coastal strand, sandhill

Mature Size (H x W): 3 ft. x 3 ft.

Growth Habit: Upright annual

Growth Rate: Fast

Description: An annual branching primarily from the base. The leaves are compound with up to 20 pairs of small leaflets. The flowers are bright yellow, often with red toward the center. The fruits are small, flat, narrow legumes about 3 inches long. It blooms most of the year.

Landscaping and Wildlife Use: This legume is grown as a soil-improving cover crop because of the nitrogen-fixing bacteria in the root nodules. Use as a mass planting. It is the "food of choice" of sulfur butterfly caterpillars. The flowers are pollinated by bumblebees. Although the seeds may be eaten in moderation by quail and other birds, they are actually toxic.

Light: Full to partial sun

Soil: Sandy, well drained

Propagation: Seeds collected before the pod shatters

Notes: Partridge pea grows throughout Florida and the eastern half of the United States.

The related ***sensitive pea (Chamaecrista nictitans)*** can also be planted, but the flowers are much smaller, making the plant less colorful.

Passionflower

Common Name: Passionflower, maypop

Botanical Name: Passiflora incarnata

Family: Passionflower (Passifloraceae)

Plant Communities: Coastal uplands, sandhill, pine flatwoods, open fields

Mature Size (H x W): 12 ft. x 5 ft.

Growth Habit: Vine

Growth Rate: Fast

Description: This winding perennial vine climbs by tendrils and may produce root suckers. The matte, green, three-lobed leaves may reach 6 inches in width, and the fringed, two-toned lavender flowers average 2 1/2 inches in width. The large greenish-yellow fruits, known as maypops, are edible.

Landscaping and Wildlife Use: Passionflower is popular with butterfly gardeners, as it is the larval plant food for zebra longwing, gulf fritillary, and Julia butterflies. Plant along a fence, arbor, or trellis; use as a groundcover; or maintain in a container. This plant may need periodic trimming as it can colonize aggressively if not controlled by caterpillars, but it does not tolerate regular mowing.

Light: Full sun

Soil: Sandy, well drained

Propagation: From seed, stem cuttings, root cuttings, layering

Notes: The name passionflower refers to the cross-shaped arrangement of flower parts, which reminds people of the crucifixion, or "passion," of Christ.

Pink Muhlygrass

Common Name: Pink muhlygrass

Botanical Name: *Muhlenbergia capillaris*

Family: Grass (Poaceae)

Plant Communities: Pine flatwoods, sandhill, coastal strand and hammocks, dunes

Mature Size (H x W): 3 1/2 ft. x 3 ft.

Growth Habit: An upright, clumping grass with an open, spreading shape

Growth Rate: Moderate

Description: Pink muhlygrass grows in clumps of wiry, rolled, gray-green leaves for most of the year. In the fall this perennial produces spectacular masses of tiny purple to pink blossoms that last 6 to 8 weeks. After blooming, the flowers turn an attractive tan color.

Landscaping and Wildlife Use: Plant pink muhlygrass in masses on roadside medians, as a backdrop for wildflower gardens, in coastal landscapes, as a groundcover spaced on 3-foot centers, and as a soil stabilizer in areas prone to erosion. Pink muhlygrass combines well with seaside goldenrod, cabbage palm, and coralbean.

Light: Full sun to light, shifting shade

Soil: Sandy, acid to alkaline

Propagation: From seed. Reseeds sparsely.

Notes: Pink muhlygrass is used by basket makers in South Carolina, who refer to it as "sweetgrass."

Pricklypear

Common Name: Pricklypear

Botanical Name: *Opuntia stricta*

Family: Cactus (Cactaceae)

Plant Communities: Beach dune, coastal strand

Mature Size (H x W): 6 ft. x 3 ft.

Growth Habit: Shrublike perennial

Growth Rate: Slow

Description: An erect, succulent, evergreen, woody-based cactus with oval, flattened, fleshy pads (stems) with 1-inch-long, yellow spines and hairlike, barbed bristles. Large (2 to 3 inches), bright yellow, 8- to 10-petaled flowers adorn the plant from spring to fall. In 2 to 3 months, the flowers are followed by an edible, pear-shaped, red fruit, actually a berry. Although common on some sites, it is a threatened species in Florida. This plant has a high salt and salt spray tolerance and tolerates infrequent inundation.

Landscaping and Wildlife Use: An excellent plant for deterring pedestrian or pet traffic! Makes an attractive specimen plant. It may be grown in a planter. Always handle with gloves as the small, barbed bristles can cause painful irritations. Excellent wildlife plant; birds including mockingbirds and catbirds eat the fruits while gopher tortoises eat the fruit and pads.

Light: Full sun

Soil: Sandy, well drained

Propagation: Break off a pad and "harden off" for a couple of days, then plant directly into well-drained soil.

Notes: Pricklypear grows in coastal areas from Virginia south throughout Florida and west to Louisiana. The fruit pulp simmered may be used to make a delicious and colorful fruit drink, syrup, jelly, or candy. The pricklypear is now threatened by a South American moth used to control this plant in Australia which escaped to Florida about 10 years ago and is already causing decline of the plant here.

The related, more inland **common pricklypear *(O. humifusa)*** has smaller pads, is usually prostrate growing, and has brown spines.

Purple Lovegrass

Common Name: Purple lovegrass

Botanical Name: Eragrostis spectabilis

Family: Grass (Poaceae)

Plant Communities: Sandhill, pine flatwoods, coastal uplands

Mature Size (H x W): 1 1/2 ft. x 1 1/2 ft.

Growth Habit: Mounding grass

Growth Rate: Fast

Description: This perennial grass forms a dense clump. Its flat blades may have surface hairs. In autumn it produces tiny, purple-red flowers that resemble a mist or haze when viewed from a distance.

Landscaping and Wildlife Use: Plant as a filler or mass planting in wildflower gardens, for plant community restoration, for erosion control, or for fall color on roadsides. It competes well with weeds and does not require mowing. Do not mulch during establishment.

Light: Full to partial sun

Soil: Sandy, acid to alkaline

Propagation: From seed. You can spread the seeds by mowing.

Notes: In the fall, this flowering grass graces many central Florida roadsides with a purple haze.

Railroad Vine

Common Name: Railroad vine, bayhops

Botanical Name: Ipomoea pes-caprae

Family: Morningglory (Convolvulaceae)

Plant Communities: Beach dune, coastal strand

Mature Size (H x W): 0.5 ft. x 75 ft.

Growth Habit: Perennial, herbaceous, trailing vine

Growth Rate: Fast

Description: Succulent vines with leathery, notched leaves to 4 inches long. The large flowers are lavender to purple. The flowers last less than 1 day. Plants may be continually flowering in southern areas but have a winter dormancy in central Florida. The large, capsular fruits contain a few large seeds.

Landscaping and Wildlife Use: Railroad vine is naturally prostrate and is a good sand/beach binder. However, it can be trailed to grow upright on a trellis or fence. Plants may be trimmed or pruned and will root at the nodes.

Light: Full sun

Soil: Sandy, alkaline, well drained

Propagation: Seeds and cuttings. Remove seed from the large, dry capsules and soak in warm water overnight before planting. Vines root at the nodes and may be cut into sections for transplanting or cut and rooted in pots or trays.

Railroad vine continued on next page

Notes: Railroad vine is found in coastal areas from the Carolinas south nearly throughout Florida and along the Gulf Coast to Texas. It occurs throughout the tropics in coastal areas. It has a notched leaf, thus the derivation of the scientific name *pes-caprae*, Latin for goat's foot. The name railroad vine refers to the plant's straight, aligned growth like that of a railroad track. The leaves are edible, although probably toxic in large amounts, as are the seeds.

Two related species are sometimes planted in coastal areas. **Beach morningglory *(I. imperati, syn. I. stolonifera)*** is a smaller plant with succulent, lobed leaves and flowers that are white with yellow or purple in the throat. It can be used like railroad vine and has high salt and salt spray tolerance. **Moonflower or tropical white morningglory *(I. alba)***, a climber, growing into tree canopies, can be grown on a trellis or fence. It has large heart-shaped or lobed leaves. The large, 3- to 4-inch-wide, funnel- or trumpet-shaped, white, fragrant flowers open only at night. Moonflower grows most lushly in shaded conditions mimicking tree canopy conditions. Beach morningglory is common in coastal areas nearly throughout Florida. Moonflower is found primarily in central and southern Florida in both coastal and inland sites. Both species occur in the Gulf Coast states with beach morningglory extending north to the Carolinas. They also occur throughout the tropics.

Rougeplant

Common Name: Rougeplant

Botanical Name: *Rivina humilis*

Family: Pokeweed (Phytolaccaceae)

Plant Communities: Maritime forest

Mature Size (H x W): 3 ft. x 2 ft.

Growth Habit: Vinelike or bushlike perennial

Growth Rate: Moderate

Description: Woody-based perennial with dark green, lance-shaped, alternating leaves and small white or pink flowers. It is the small, bright red berries that make this a popular landscaping plant. It flowers most of the year.

Landscaping and Wildlife Use: Its dark green leaves and dense vinelike growth, coupled with the red berries, make this an attractive plant for naturalizing or mass planting in full sun to partial shade. It has medium salt tolerance.

Light: Full sun to shade

Soil: Sandy, well drained; tolerates some moisture and does best in an organic soil

Propagation: Self-seeding, or plant seeds 1/4 inch deep after removing from berry. Plant in a moist, sandy soil for germination in 1 to 2 weeks. Easily transplanted.

Notes: Rougeplant grows throughout Florida and west to Arizona. It also occurs in the West Indies, Mexico, Central America, and South America. The berries were used as a dye by Native Americans.

Sand Cordgrass

Common Name: Sand cordgrass

Botanical Name: *Spartina bakeri*

Family: Grass (Poaceae)

Plant Communities: Salt- and freshwater marshes

Mature Size (H x W): 6 ft. x 4 ft.

Growth Habit: Erect bunchgrass

Growth Rate: Fast

Description: Sand cordgrass forms large, evergreen, fountainlike clumps with leaf blades up to 6 ft. long that are involute (rolled inward from the edges) and rough to the touch. The green inflorescences reach to 6 ft. in height.

Landscape and Wildlife Use: An attractive substitute for the nonnative pampas grass. Plant sand cordgrass clumps at least 3 to 6 feet apart. It is a good soil stabilizer, often used on slopes of wetland mitigation sites. Wildlife use the large clumps for nesting and cover. The seeds are eaten by ducks, marsh birds, and song birds. It is surprisingly drought tolerant and can be used anywhere in the landscape, even in sandy, well-drained situations.

Light: Full sun to partial shade

Soil: Sandy, wide range of pH

Propagation: Division

Notes: Sand cordgrass occurs in coastal South Carolina, Georgia, nearly throughout Florida, and in Texas. It is the only *Spartina* species that is naturally occurring in freshwater communities.

Seaoats

Common Name: Seaoats

Botanical Name: Uniola paniculata

Family: Grass (Poaceae)

Plant Communities: Beach dune, coastal strand

Mature Size (H x W): 6 ft. x 3 ft.

Growth Habit: Sparse perennial running grass

Growth Rate: Fast

Description: Seaoats is a coarse perennial, evergreen grass with extensive creeping rhizomes (underground runners) that root readily at the nodes. The leaves emerge at the nodes and are 36 to 40 inches long. The green inflorescence is a panicle with a flattened flower structure that resembles oats.

Landscape and Wildlife Use: Seaoats is an excellent dune stabilizer and source of food for birds and small mammals. Use it in a beachside planting with railroad vine and dune sunflower. Remove the dead growth, which is thought to harbor fungal diseases. Seaoats will die to the ground in cold winters. Plant from March through November.

Light: Full sun

Soil: Sandy, wide range of pH

Seaoats continued on next page

Propagation: The seeds have low viability. Divide mature clumps and reestablish in pots before planting.

Notes: Seaoats ranges from Virginia south to Florida and west to Texas. It also occurs in the West Indies and Mexico. It is unlawful to gather any part of seaoats in many states. It was the first plant protected by law in Florida. The seeds are edible, and the young plants are favored by marsh rabbits, which can destroy plants. Wire cages are sometimes used to protect seaoats from rabbits.

Groundcovers, Vines, and Wildflowers

Seashore Paspalum

Common Name: Seashore paspalum

Botanical Name: *Paspalum vaginatum*

Family: Grass (Poaceae)

Plant Communities: Beach dune, salt marsh

Mature Size (H): 3 ft.

Growth Habit: Perennial runner grass forming extensive dense mats

Growth Rate: Fast

Description: Seashore paspalum is a perennial, evergreen grass with decumbent stems (runners) that root at the nodes and with erect flowering stems. The leaves are 1 to 6 inches long and 1/8 to 1/4 inch wide and are mostly glabrous. Loose sheaths (lower part of the leaf that encloses the stem) differentiate this from other similar appearing coastal grasses. The green inflorescence is two- to sometimes three-branched.

Landscape and Wildlife Use: Treat this grass like Bahiagrass, mowing on the highest mower setting for a low dense mat. It is a good soil stabilizer and is a source of food and cover for birds and small mammals if left to grow tall. Plant from March to November.

Light: Full sun

Soil: Sandy to mucky, wide range of pH

Propagation: Division, seeds

Notes: Seashore paspalum occurs along the coast from North Carolina south throughout Florida and west to Texas. It is a common grass in coastal areas throughout the tropics. It is being increasingly used as a turf grass in coastal areas because of its high salt tolerance.

Seaside Goldenrod

Common Name: Seaside goldenrod

Botanical Name: *Solidago sempervirens*

Family: Aster (Asteraceae)

Plant Communities: Beach dune, coastal strand, salt marsh

Mature Size (H x W): 6 ft. x 1.5 ft.

Growth Habit: Perennial, herbaceous wildflower

Growth Rate: Fast

Description: This unbranched plant has long, wide, succulent basal leaves and a tall flowering stem with smaller, alternating leaves on the stem. It has a club-shaped panicle of yellow flowers in small heads with straplike ray flowers and tubular disc flowers. The plants bloom almost year-round with the heaviest bloom in the late summer and fall.

Landscaping and Wildlife Use: Perfect for the back of a border where they go virtually unnoticed until they send up the magnificent flower stem in late summer or fall, adding a new dimension to the border. Monarch butterflies feed on the nectar from the flowers in the fall.

Light: Full to partial sun

Soil: Sandy, dry to moist

Propagation: Seeds, often self-seeding; division of rhizomes

Notes: Seaside goldenrod occurs in coastal areas from Canada south throughout Florida and west to Texas. It also occurs in the West Indies, Mexico, Central America, and South America. These plants are not harbingers of hay fever as previously believed; they just have the misfortune of blooming at the same time as the guilty ragweed.

Spotted Beebalm

Common Name: Spotted beebalm, dotted horsemint

Botanical Name: Monarda punctata

Family: Mint (Lamiaceae)

Plant Communities: Coastal strand, sandhill

Mature Size (H x W): 3 ft. x 2.5 ft.

Growth Habit: Perennial subshrubby wildflower

Growth Rate: Fast

Description: The opposite leaves are lance-shaped and toothed. The purple-spotted, 2-lipped, yellow flower clusters are whorled around the hairy square stems. Green and pale lavender to cream-colored bracts occur at the base of each cluster, becoming the most conspicuous part of the inflorescence. It blooms late spring through fall.

Landscaping and Wildlife Use: This shrubby perennial goes dormant in the winter and should be severely pruned while dormant. Use in mass plantings in areas where nectar-feeding hummingbirds and butterflies may be easily observed. It has medium salt tolerance.

Light: Full to partial sun

Soil: Sandy, well drained, can tolerate some moisture

Propagation: The plant frequently reseeds itself. Seeds may also be collected after the outermost petals turn dark brown in late summer about three weeks after flowering. Test by bending the stem over and tapping to see if seeds drop freely. Snip off seed heads, place in a paper bag, and shake vigorously. Store seeds in an airtight container in the refrigerator until planting in early spring, at which time they will germinate in one or two weeks. Take 3- to 4-inch stem tip cuttings in the spring or summer. Mature clumps can be divided in the winter while the plant is dormant.

Notes: This member of the mint family grows throughout most of Florida and is common in most of eastern United States and the Midwest to Minnesota and New Mexico. It also occurs in California. It has aromatic leaves that make a good tea. Prepare it as you would oriental tea. It also is a good plant for flower arrangements.

Tropical Sage

Common Name: Tropical sage

Botanical Name: Salvia coccinea

Family: Mint (Laminaceae)

Plant Communities: Shell mounds, hammocks, disturbed thickets

Mature Size (H x W): 4 ft. x 2 ft.

Growth Habit: Upright wildflower

Growth Rate: Fast

Description: This tall perennial produces brilliant scarlet, two-lipped flowers during most of the year. The toothed leaves grow directly across from each other, and the stems are hairy. Tropical sage is intolerant of freezing temperatures but recovers quickly, usually growing back from the roots and always reseeding. Each plant yields many seedlings.

Landscaping and Wildlife Use: Plant in wildflower gardens and as a nectar source for both butterflies and hummingbirds. It combines well with yellow, purple, or blue flowers, with green foliage, and with palms.

Light: Full sun

Soil: Sandy, well drained

Propagation: From seed, stem cuttings

Notes: Nurseries sell white and pink varieties.

QUICK REFERENCE GUIDE

Trees

Common Name *Scientific Name*	**Page No.**	**Coastal Plant Community**	**Evergreen**	**Growth Rate**
American holly *Ilex opaca*	12	Mf	Yes	Moderate
Live oak *Quercus virginiana*	14	Cs	Yes	Moderate
Myrtle oak *Quercus myrtifolia*	16	Bd, Cs	Yes	Moderate
Persimmon *Diospyros virginiana*	18	Mf	No	Moderate
Red bay *Persea borbonia*	20	Mf	Yes	Moderate
Redbud *Cercis canadensis*	22	Mf	No	Moderate
Red cedar *Juniperus virginiana*	24	Cs, Mf	Yes	Moderate
Sand live oak *Quercus geminata*	26	Cs, Mf	Yes	Moderate
Slash pine *Pinus elliottii*	28	Mf	Yes	Moderate
Southern magnolia *Magnolia grandiflora*	30	Mf	Yes	Moderate
Sugarberry *Celtis laevigata*	34	Mf	No	Moderate–fast
Sweet acacia *Acacia farnesiana*	36	Cs, Mf	No	Slow–moderate

Key to plant communities (described on pages 139–140)
Bd = Beach dune   Cs = Coastal strand   Mf = Maritime forest

Mature Size (H x W)	Flower Color	Showy Season	Shade Tolerant	Moisture Tolerant
40' x 30'	White	All year	Partial	No
80' x 100'	—	All year	Partial	Yes
40' x 8'	—	All year	Partial	Yes
50' x 40'	Greenish yellow	Spring/fall	Partial	No
50' x 40'	Greenish yellow	Spring/fall	Partial	No
30' x 20'	Lavender-pink	Spring/fall	Partial	No
90' x 25'	—	All year	No	No
30' x 20'	—	All year	No	No
60' x 40'	—	All year	No	No
75' x 35'	White	All year	No	Yes
80' x 40'	Green	Spring/fall	No	Yes
15' x 15'	Yellow	Spring/fall	No	No

— = not applicable

Shrubs

Common Name *Scientific Name*	**Page No.**	**Coastal Plant Community**	**Evergreen**	**Growth Rate**
Adam's Needle *Yucca filamentosa*	40	—	Yes	Moderate
Bay cedar *Suriana maritima*	42	Bd, Cs	No	Slow
Beautyberry *Callicarpa americana*	44	Cs, Mf	No	Fast
Christmasberry *Lycium carolinianum*	46	Bd, Cs	No	Slow
Coontie *Zamia pumila*	48	—	Yes	Slow
Coralbean *Erythrina herbacea*	50	Cs	No	Fast
Firebush *Hamelia patens*	52	Cs, Mf	No	Fast
Florida privet *Forestiera segregata*	54	Bd, Cs, Mf	Yes	Moderate
Inkberry *Scaevola plumieri*	56	Bd, Cs	Yes	Slow–moderate
Marlberry *Ardisia escallonioides*	58	Mf	Yes	Moderate
Necklacepod *Sophora tomentosa*	60	Bd, Cs	Yes	Slow
Seagrape *Coccoloba uvifera*	62	Bd, Cs, Mf	Yes	Moderate
Simpson's stopper *Myrcianthes fragrans*	64	Mf	Yes	Moderate
Snowberry *Chiococca alba*	66	Mf	Yes	Slow–moderate
Sparkleberry *Vaccinium arboreum*	68	Mf	No	Slow

Key to plant communities (described on pages 139–140)
Bd = Beach dune Cs = Coastal strand Mf = Maritime forest

Mature Size (H x W)	Flower Color	Showy Season	Shade Tolerant	Moisture Tolerant
2' x 2'	White	Spring/summer	No	No
20' x 10'	Yellow	All year	No	No
7' x 6'	Lavender-pink	Spring/fall	Yes	Yes
9' x 3'	Blue, lavender, or white	Spring/fall	Partial	Yes
2' x 3'	—	All year	Partial	Yes
16' x 4'	Red	Spring	Partial	Yes
12' x 8'	Red-orange	Spring/fall	Partial	Yes
20' x 10'	Greenish yellow	All year	Partial	Yes
6' x 6'	White or pinkish	All year	No	No
15' x 5'	White-pink	All year	Yes	Yes
10' x 6'	Yellow	All year	No	No
50' x 20'	Yellow-green	All year	Partial	Yes
20' x 6'	White	All year	Yes	Yes
10' x 10'	Yellow	All year	Yes	Yes
30' x 15'	White	Spring/fall	Partial	No

— = not applicable

Common Name *Scientific Name*	Page No.	Coastal Plant Community	Evergreen	Growth Rate
Tough Bully *Sideroxylon tenax*	70	Cs, Mf	Yes	Slow
Varnishleaf *Dodonaea viscosa*	72	Bd, Cs	Yes	Slow–moderate
Wild coffee *Psychotria nervosa*	74	Mf	Yes	Moderate

Groundcovers, Vines, and Wildflowers

Common Name *Scientific Name*	Page No.	Coastal Plant Community	Evergreen	Growth Rate
Beach verbena *Glandularia maritima*	78	Bd, Cs, Mf	No	Fast
Blanketflower *Gaillardia pulchella*	80	Bd, Cs	No	Fast
Bluecurls *Trichostema dichotomum*	82	Mf	No	Fast
Blue porterweed *Stachytarpheta jamaicensis*	84	Cs, Mf	No	Fast
Corkystem passionflower *Passiflora suberosa*	86	Mf	No	Moderate
Dayflower *Commelina erecta*	88	Bd, Cs	No	Moderate
Dune sunflower *Helianthus debilis*	90	Bd, Cs	No	Fast
Dwarf Fakahatcheegrass *Tripsacum floridanum*	94	—	Yes	Fast

Key to plant communities (described on pages 139–140)

Bd = Beach dune   Cs = Coastal strand   Mf = Maritime forest

Mature Size (H x W)	Flower Color	Showy Season	Shade Tolerant	Moisture Tolerant
20' x 15'	White	All year	Partial	No
15' x 6'	Yellowish green	All year	No	No
10' x 3'	White	All year	Yes	Yes

Mature Size (H x W)	Flower Color	Showy Season	Shade Tolerant	Moisture Tolerant
1' x 3'	Rose or purple	Spring/fall	No	No
1' x 2'	Yellow, red, and purple	Spring/fall	No	No
3' x 3'	Blue	Spring/fall	Partial	No
2' x 3'	Blue	Spring/fall	Partial	Yes
0.5' x 20'	Greenish yellow	Spring/fall	Partial	Yes
2' x 1'	Blue	Spring/fall	No	No
3' x 6'	Yellow & brown	Spring/fall	No	No
5' x 4'	Green	All year	No	Yes

— = not applicable

Common Name *Scientific Name*	**Page No.**	**Coastal Plant Community**	**Evergreen**	**Growth Rate**
Marshhay cordgrass *Spartina patens*	96	Bd	Yes	Fast
Partridge pea *Chamaecrista fasciculata*	98	Cs	No	Fast
Passionflower *Passiflora incarnata*	100	—	No	Fast
Pink Muhlygrass *Muhlenbergia capillaris*	102	Cs	No	Moderate
Pricklypear *Opuntia stricta*	104	Bd, Cs	Yes	Slow
Purple lovegrass *Eragrostis spectabilis*	106	—	Yes	Fast
Railroad vine *Ipomoea pes-caprae*	108	Bd, Cs	No	Fast
Rougeplant *Rivina humilis*	112	Mf	No	Moderate
Sand cordgrass *Spartina bakeri*	114	—	Yes	Fast
Seaoats *Uniola paniculata*	116	Bd, Cs	Yes	Fast
Seashore paspalum *Paspalum vaginatum*	120	Bd	Yes	Fast
Seaside goldenrod *Solidago sempervirens*	122	Bd, Cs	No	Fast
Spotted beebalm *Monarda punctata*	124	Cs	No	Fast
Tropical sage *Salvia coccinea*	126	—	No	Fast

Key to plant communities (described on pages 139–140)
Bd = Beach dune Cs = Coastal strand Mf = Maritime forest

Mature Size (H x W)	Flower Color	Showy Season	Shade Tolerant	Moisture Tolerant
3' high	Green	All year	No	Yes
3' x 3'	Yellow	Spring/fall	Partial	No
12' x 5'	Lavender	Summer	No	No
3' x 3'	Purple to pink	Fall	Partial	Yes
6' x 3'	Yellow	All year	No	No
2' x 2'	Purple to reddish	Fall	Partial	No
0.5' x 75'	Lavender-purple	Spring/summer	No	Yes
3' x 2'	White or pink	Spring/fall	Yes	Yes
6' x 4'	Green	All year	Partial	Yes
6' x 3'	Green	All year	No	No
3' high	Green	All year	No	Yes
6' x 1.5'	Yellow	Summer/fall	Partial	Yes
3' x 2.5'	Lavender-cream	Spring/fall	Partial	Yes
4' x 2'	Red	Spring/fall	No	Yes

— = not applicable

OTHER FAVORITE SALT-TOLERANT NATIVES FOR DRY SITES

The following are just a few of the species we had to leave out for lack of space. You will find many of these and numerous other salt-tolerant dry-site plants described in the publications listed in the bibliography.

Common Name	Botanical Name	Plant Type
Cabbage palm	*Sabal palmetto*	Tree
Chickasaw plum	*Prunus angustifolia*	Tree
Seaoxeye	*Borrichia frutescens*	Wildflower
Sea purslane	*Sesuvium portulacastrum*	Groundcover
Wax myrtle	*Myrica cerifera*	Shrub
Wild sage, button sage	*Lantana involucrata*	Groundcover
Yaupon holly	*Ilex vomitoria*	Shrub
Yellowtop	*Flaveria linearis*	Wildflower

RESOURCES

Florida Native Plant Society (FNPS)

Your local chapter of the FNPS is an excellent source of information about gardening with native plants. FNPS members enjoy sharing their experiences, their expertise, and their plants. Informative and enjoyable field trips, meetings, presentations, newsletters, and libraries all contribute to the knowledge and enthusiasm of members.

The FNPS publishes a quarterly magazine, *The Palmetto*, the official magazine of the society. *The Palmetto* contains articles about native and exotic plants, native plant landscaping, backyard habitats, and other topics related to native plant conservation and education. Most FNPS chapters also produce newsletters for their local membership. For example, the Suncoast Native Plant Society publishes a monthly newsletter, *The Suncoast Grapevine*, providing articles about upcoming activities, conservation, native plant landscaping, wildlife, and interesting places to visit. For the name of the chapter nearest you, contact the Florida Native Plant Society at P.O. Box 278, Melbourne, FL 32902-0278, or visit the FNPS website at www.fnps.org.

County Cooperative Extension Service

Your local county Cooperative Extension Service can answer many of your questions as a free public service paid for by your taxes. Some particularly helpful and interesting Extension services designed to foster conservation of natural resources are Florida Friendly Landscaping, Florida Master Gardener, Florida Master Naturalist, Florida Wildlife, and Florida Forest Stewardship programs. Visit www.solutionsforyourlife.ufl.edu for detailed information about these programs and for contact information for your local Extension Service office.

Water Management Districts

Water Management Districts are good sources of information about gardening with native plants and water conservation. The Water Management Districts that serve peninsular Florida are:

St. Johns River Water Management District (www.floridaswater.com, telephone 800-451-7106) for east-central Florida

Southwest Florida Water Management District (www.swfwmd.state.fl.us, telephone 800-423-1476) for west-central Florida

South Florida Water Management District (www.sfwmd.gov, telephone 800-432-2045) for southern Florida

Places to See Native Plants

Florida has many protected or managed natural areas for viewing native plants and the habitats that native plants support. The U.S. government maintains national forests, parks, monuments, and recreation areas for viewing native plants. The Florida State Park system includes many natural areas, while counties maintain parks, preserves, and environmentally sensitive lands for viewing native plants. Many of these locations provide interpretative centers and self-guided tours.

Native Plant Sources

Because of concern about the availability of water resources, many retail nurseries stock a variety of native plants. Many nursery personnel will gladly special order plants for you.

Local FNPS chapter members are excellent sources of knowledge about local nurseries with good selections of native plants. Some wholesale nurseries will sell to the public for a minimum order or to members of the FNPS as a chapter field trip.

The Florida Association of Native Nurseries (www.afnn.org) is an excellent resource for finding local native plants.

DRY PENINSULAR FLORIDA PLANT COMMUNITIES

Three major coastal plant communities occur in central Florida (in order from the water's edge to upland): beach dune, coastal strand, and maritime forest. If your site meets the criteria under "Who Needs Salt-Tolerant Plants for Dry Areas?" on page 2, most likely it originally supported one of these communities. However, you do not need to know which one, because any of the plants listed in this book can grow on your dry site, as long as the light requirements are met. The following descriptions characterize each community.

Beach Dune

This community is characterized as a wind-deposited foredune and wave-deposited upper beach that are sparsely to densely vegetated with pioneer species such as seaoats, railroad vine, beach morning-glory, seashore paspalum, and dune sunflower. The character of the beach dune is dominated by the physical processes of wind and waves that cause its formation and (sometimes) destruction. The beach dune is usually only lightly vegetated due to storm wash and wind-induced changes to the sandy surface, making establishment difficult. Plant species that are established in the coastal strand may colonize the beach dune until the next bout of wind and wave damage destroys them.

Coastal Strand

Coastal strands are built up by the action of onshore winds sweeping grains of sand from the beach into the area where plants grow upward. Plants cause the wind speed to slow and result in the grains of sand being deposited on the downwind (inland) side of the plants. Seaoats (*Uniola paniculata*) is the most important in dune building in north and central Florida. As the dune grows, other species can become established. All of these species are very salt tolerant. Seaoats, other grasses and sedges, and low scrubby plants will grow on the windward side of the dune. The lee side of the dune will feature scrubby shrubs and trees, mostly of a uniform maximum height due to salt spray pruning. Typical species of the coastal strand are saw palmetto, sand live oak, cabbage palm, yaupon holly, seagrape, bay cedar, and necklacepod.

Maritime Forest

The maritime forest is the upland region characterized as a narrow band of hardwood forest starting at the inland side of the coastal strand. It is not directly affected by the dune building processes. The maritime forest receives wind-blown salt spray, although in much smaller quantities than do the beach dunes and coastal strand. Plant species in the maritime forest may be somewhat less salt tolerant than those species occurring closer to the water. There will also be a recognizable canopy and understory. Typical plant species include American holly, Southern magnolia, red cedar, seagrape, saw palmetto, beautyberry, coralbean, coontie, wild coffee, snowberry, marlberry, and rouge plant.

FLORIDA NATIVE PLANT SOCIETY POLICY ON TRANSPLANTING NATIVE PLANTS FROM THE WILD

(adopted by FNPS Board of Directors, Nov. 19, 1988)

By promoting the use of native plants in landscaping, the Florida Native Plant Society has helped create a demand for native plants. This demand has prompted some individuals and companies to offer for sale native plants dug from the wild. Selling harvested plants avoids the time and expense of growing the plants under nursery conditions, but this harvest of native plants is, in most cases, not sustainable in the long run, and damages land that has potential for preservation.

FNPS does not approve of transplanting native plants from natural areas for landscaping, mitigation, or restoration purposes. Such transplanting is in direct conflict with the Society's goal of preservation and conservation of native plants in their natural habitats.

From its beginning, the Society has promoted the preservation of existing native plants in their natural habitats, while encouraging the planting of additional native plants grown under nursery conditions from seeds or cuttings. This position should result in a net gain of native plants. Transplanting from the wild merely transfers native plants from natural to manmade habitats, and results in a net loss of native plants because a good percentage of transplanted plants fail to survive the transplanting process.

The society recognizes that there should be an exception to this policy to allow the salvage of native plants from areas where land clearing activities are both imminent and assured. This type of transplanting saves plants that would otherwise be lost. It is expected, however, that salvage activities not take place until all planning approvals for the site have been obtained and all possibilities for preservation have been exhausted. Salvage and plant rescue operations should be undertaken only in compliance with all state and local native plant protection laws.

Members of the society are asked to abide by this policy as a matter of ethics. Specifically, members are asked to inquire about the origins of plant material and not buy plants that have been transplanted from the wild. Landscape architects and designers are asked, when writing plant material specifications, to specify only nursery-grown native plants. Government agencies with jurisdiction over landscaping, mitigation, and restoration projects are asked to require those projects to use only nursery-grown native plants or those from on-site or nearby salvage operations.

While it is unrealistic to think that FNPS can totally stop the practice of transplanting from the wild, it can supply needed leadership on this issue and, with the support of its members, help dry up the market for such plants. It is critical that native plant communities remain as undisturbed and undamaged as possible.

BIBLIOGRAPHY

Books marked with an asterisk (*) are especially helpful for the beginning native plant gardener.

*Austin, Daniel. 1993. *Scrub Plant Guide: A Pocket Guide to the Common Plants of Southern Florida's Scrub Community.* Boca Raton, FL: Gumbo Limbo Nature Center of South Palm Beach County.

*Barnett, Michael R., and David W. Crewz, eds. 1997. *Common Coastal Plants in Florida*. Gainesville, FL: University Press of Florida.

*Bell, C. Ritchie, and Bryan J. Taylor. 1982. *Florida Wild Flowers and Roadside Plants.* Chapel Hill, NC: Laurel Hill Press.

*Bir, Richard E. 1992. *Growing and Propagating Showy Native Woody Plants.* Chapel Hill, NC: The University of North Carolina Press.

Bowman, Sheryl, Debbie Butts, Betsy Davis, George Kish, John Marsh, Ann Nord, Carl Strohmenger, and Richard Wunderlin. 2005. *The Right Plants for Dry Places: Native Plant Landscaping in Central Florida,* Second Edition. St. Petersburg, FL: Great Outdoors Publishing Co.

*Cerulean, Susan, Celeste Botha, and Donna Legare. 1987. *Planting a Refuge for Wildlife: How to Create a Backyard Habitat for Florida's Birds and Beasts.* Tallahassee, FL: Florida Game and Fresh Water Fish Commission Nongame Wildlife Program, United States Department of Agriculture Soil Conservation Service.

*Craig, Robert M. 1991. *Plants for Coastal Dunes of the Gulf and South Atlantic Coasts and Puerto Rico*. Gainesville, FL: Soil Conservation Services, Agricultural Information Bulletin 460.

*Foote, Leonard E., and Samuel B. Jones. 1989. *Native Shrubs and Woody Vines of the Southeast.* Portland, OR: Timber Press.

*Haehle, Robert G., and Joan Brookwell. 2004. *Native Florida Plants,* Revised Edition. Lanham, MD: Taylor Trade Publishing Company.

*Hall, David W. 1993. *Illustrated Plants of Florida and the Coastal Plain.* Gainesville, FL: Maupin House Publishing.

Hartman, Hudson T., Dale E. Kester, Fred T. Davies, and Robert L. Geneve. 2010. *Plant Propagation: Principles and Practice,* 8th Edition. Upper Saddle River, NJ: Prentice Hall.

*Huegel, Craig N. 2010. *Native Plant Landscaping for Florida Wildlife.* Gainesville, FL: University Press of Florida.

*Jameson, Michael, and Richard Moyroud, eds. 1991. *Xeric Landscaping with Florida Native Plants.* Hollywood, FL: Betrock Information Systems.

Myers, Ronald L., and John J. Ewel, eds. 1990. *Ecosystems of Florida.* Orlando, FL: University of Central Florida Press.

Nellis, David. 1994. *Seashore Plants of South Florida and the Caribbean: A Guide to Knowing and Growing Drought- and Salt-Tolerant Plants*. Sarasota, FL: Pineapple Press.

*Nelson, Gil. 2010. *Best Native Plants for Southern Gardens.* Gainesville, FL: University Press of Florida.

*Nelson, Gil. 2003. *Florida's Best Native Landscape Plants: 200 Readily Available Species for Homeowners and Professionals*. Gainesville, FL: University Press of Florida.

*Nelson, Gil. 1996. *The Shrubs and Woody Vines of Florida: A Reference and Field Guide*. Sarasota, FL: Pineapple Press.

*Nelson, Gil. 2011. *The Trees of Florida: A Reference and Field Guide,* Second Edition. Sarasota, FL: Pineapple Press.

*Osorio, Rufino. 2001. *A Gardener's Guide to Florida Native Plants.* Gainesville, FL: University Press of Florida.

*Phillips, Harry R. 1985. *Growing and Propagating Wild Flowers.* Chapel Hill, NC: The University of North Carolina Press.

*Suncoast Native Plant Society. 1997. *The Right Plants for Dry Places.* St. Petersburg, FL: Great Outdoors Publishing Company.

*Taylor, Walter Kingsley. 1998. *Florida Wildflowers in Their Natural Communities.* Gainesville, FL: University Press of Florida.

*Taylor, Walter Kingsley. 2009. *A Guide to Florida Grasses.* Gainesville, FL: University Press of Florida.

*Taylor, Walter Kingsley. 1992. *The Guide to Florida Wildflowers.* Dallas, TX: Taylor Publishing Company.

*Traas, Pamela F. 1999. *Gardening for Florida's Butterflies.* St. Petersburg, FL: Great Outdoors Publishing Company.

Walton, Dan, and Laurel Schiller. 2007. *Natural Florida Landscaping.* Sarasota, FL: Pineapple Press.

*Wasowski, Sally, and Andy Wasowski. 1994. *Gardening with Native Plants of the South.* Dallas, TX: Taylor Publishing Company.

*Workman, Richard W. 1980. *Growing Native: Native Plants for Landscape Use in Coastal South Florida.* Sanibel, FL: The Sanibel-Captiva Conservation Foundation.

Wunderlin, Richard P., and Bruce F. Hansen. 2000. *Flora of Florida,* Volume 1: *Pteridophytes and Gymnosperms.* Gainesville, FL: University Press of Florida.

Wunderlin, Richard P., and Bruce F. Hansen. 2011. *Guide to the Vascular Plants of Florida,* Third Edition. Gainesville, FL: University Press of Florida.

PHOTO CREDITS

Jan Allyn: back cover (map), iii (map)

Deborah J. Chayet, Florida Native Plant Society, Pinellas Chapter: 101, 103, 107

June Cussen: 49 (right)

Shirley Denton: front cover (all except bottom right), 19, 21 (lower right), 23 (upper right), 25 (upper and lower right), 27, 29, 32, 33, 35, 37 (lower left, upper and lower right), 43, 45 (upper and lower left), 47, 53 (left and lower right), 55, 57, 59, 61, 63 (upper left), 65 (upper left), 67, 69 (inset), 71 (right and inset), 73, 75 (lower left and right), 81, 83, 85 (upper left and right), 87, 91, 93, 95 (right), 97, 99, 105 (upper and lower left), 111, 113, 115 (top), 118, 119, 123, 125

Joel Jackson: front cover (bottom right), 89 (upper and lower left)

George R. Kish, St. Petersburg College: 37 (upper left), 65 (right), 75 (upper left), 79, 85 (lower left), 95 (lower left), 105 (right), 115 (bottom)

Matthew Kish, 37th Age Productions: 21 (upper right), 41, 49 (inset), 51, 63 (lower left), 89 (right), 127

Robert Upcavage: title page, 15, 17, 21 (left), 23 (lower right), 25 (left), 45 (right), 65 (lower left), 69 (top), 71 (upper and lower left), 95 (upper left), 121

© Afonskaya | Dreamstime.com: 23 (left)

© Arkorn | Dreamstime.com: 117

© Astrocady | Dreamstime.com: 63 (right)

© Marzolino | Dreamstime.com: 31

© Mouce101 | Dreamstime.com: 109

© Photobytan... | Dreamstime.com: 53 (upper right)

© Wordplanet | Dreamstime.com: 13

INDEX OF COMMON NAMES

INDEX OF BOTANICAL NAMES

Here are some other books from Pineapple Press on related topics. For a complete catalog, visit our website at www.pineapplepress.com. Or write to Pineapple Press, P.O. Box 3889, Sarasota, Florida 34230-3889, or call (800) 746-3275.

Southern Gardening: An Environmentally Sensitive Approach by Marie Harrison. A comprehensive guide to beautiful, environmentally conscious yards and gardens. Suggests useful groundcovers and easy-care, adaptable trees, shrubs, perennials, and annuals.

Gardening in the Coastal South by Marie Harrison. A master gardener discusses topics such as salt tolerance, pesticide use, beneficial insects, invasive exotics, and gardening for butterflies and birds. Color photos and pen-and-ink illustrations round out the text.

Groundcovers for the South by Marie Harrison. Presents a variety of plants that can serve as groundcovers in the American South. Each entry gives detailed information on ideal growing conditions, plant care, and different selections within each species. Color photographs and line drawings make identification easy.

Flowering Shrubs and Small Trees for the South by Marie Harrison. Master gardener Marie Harrison offers tips on how to identify, select, and care for more than 100 flowering shrubs and small trees suited to the South. Full-color photos and line drawings throughout.

Natural Florida Landscaping by Dan Walton and Laurel Schiller. This book will help you create a plan that will work for your yard and choose the native plants that will thrive there in order to create a beautiful and environmentally sensitive landscape.

The Art of South Florida Gardening, Second Edition, by Harold Songdahl and Coralee Leon. Gardening advice specifically written for the unique conditions of south Florida. This practical, comprehensive guide, written with humor and know-how, will teach you how to outsmart the soil, protect against pests and weather, and select the right trees and shrubs for Florida's climate.

Flowering Trees of Florida by Mark Stebbins. Written for both the seasoned arborist and the weekend gardener alike, this comprehensive guide offers 74 outstanding tropical flowering trees that will grow in Florida's subtropical climate. Full-color photos throughout.

Ornamental Tropical Shrubs by Amanda Jarrett. Eighty-three shrubs that display beautiful flowers and/or leaves in the subtropical and tropical zones. This book includes a full profile for each shrub, including drought and salt tolerance, flowers and fruits, potential problems, and much more. Explains how to use shrubs in garden design. Full-color photos throughout.

CPSIA information can be obtained at www.ICGtesting.com
Printed in the USA
BVOW10s1102031013

332817BV00003B/3/P

9 781561 645602